D0753413

DID TIME
BEGIN?

WILL TIME
END?

DID TIME
BEGIN?

WILL TIME
END?

Paul H Frampton
University of North Carolina, USA

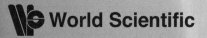 World Scientific

NEW JERSEY · LONDON · SINGAPORE · BEIJING · SHANGHAI · HONG KONG · TAIPEI · CHENNAI

Published by

World Scientific Publishing Co. Pte. Ltd.

5 Toh Tuck Link, Singapore 596224

USA office: 27 Warren Street, Suite 401-402, Hackensack, NJ 07601

UK office: 57 Shelton Street, Covent Garden, London WC2H 9HE

British Library Cataloguing-in-Publication Data
A catalogue record for this book is available from the British Library.

DID TIME BEGIN? WILL TIME END?
Maybe the Big Bang Never Occurred

ISBN-13 978-981-4280-58-7
ISBN-10 981-4280-58-5

Typeset by Stallion Press
Email: enquiries@stallionpress.com

Printed in Singapore by World Scientific Printers

Preface

Did time begin at a Big Bang? Will the present expansion of the universe last for a finite or infinite time? These questions sound philosophical but are now becoming in the twenty-first century central to the scientific study of cosmology. The answers, which should become clearer in the next decade or two, could have profound implications for how we see our own role in the universe. Since the publication of Stephen Hawking's *A Brief History of Time* in 1988, the answers to these questions have progressed as a result of research by a community of active theoretical physicists including myself. To present the underlying ideas requires a discussion of a wide range of topics in cosmology, especially the makeup of the energy content of the universe. A brief summary of my conclusions, that of three different possibilities concerning the history and future of time, the least likely being the conventional wisdom (time began and will never end) and most likely being a cyclic model (time never begins or ends), is in the short final chapter which could be read first. To understand the reasoning leading to those conclusions could encourage reading of the entire book. My hope in writing this, my first popular book, is that it will engender reflection about time. Many nonscientists may already hold a philosophical opinion about whether time begins and ends. This book aims to present some recently discovered scientific facts which can focus the reader's consideration of the two brief questions in the book title.

Paul H. Frampton
University of North Carolina at Chapel Hill

Contents

Why do Many Other Scientists Believe Time Began at a Big Bang?

Our everyday perception of the universe comes from looking up at the sky to see the Sun in the daytime and, more profoundly, to see thousands of stars in the night sky. Surely some of the oldest questions since the beginnings of human thought are: How large is the universe? Did it ever begin? What are the principal constituents of the present universe? Will time ever end?

Cosmology is the name for the scientific study of the universe. The present time is an unprecedented age for cosmology because it is fair to say that in the last twenty years we have learned more in cosmology than in all of previous human history. Despite this enormous and exciting growth of our knowledge as a result of many impressive observations, the universe has become more enigmatic in many ways. The more we learn, the more the extent of our ignorance becomes manifest.

Cosmology has recently answered some of the old questions and in this chapter we shall give answers to the first two: How large is the universe? How long ago did it — at least the present expansion era — begin? We can all agree that the expansion stage we are presently in began a finite time ago but, as I shall explain later in this chapter, it is not obvious that time itself began then, if ever.

We do know how large the visible universe is, meaning how far away the most distant galaxies are whose light can reach us on the Earth. It is theoretically possible, and even favored in some theoretical scenarios, that our universe is actually much larger than the visible universe. In some very speculative scenarios the universe is spatially finite with non-trivial topology. This is at present not readily testable so we shall be content to try to convey just how gigantic the visible part is.

The observational means by which we know accurately the size of the universe will not concern us here but it is sufficient to say that present studies using the Hubble Space Telescope combined with the largest (up to 10 meters in diameter) ground-based optical telescopes tell us the size of the visible galaxy to an accuracy of a few per cent. This sort of accuracy has been achieved only since the turn of the 21st century.

ॐ ॐ ॐ ॐ

Cosmological distances are so much bigger than any distance with which we may be familiar, it is not easy to grasp or comprehend them even in our imagination. So let us begin with the largest distance which is easily comprehensible from the viewpoint of our experience.

A very long airplane ride may take 15 hours and go 9000 miles, a significant part of halfway around the Earth. People who travel a lot may take such a flight a few times each year. One knows that the plane has a ground speed of about 600 miles per hour and the discomfort of sitting, especially in economy class, for such a long time gives a strong impression of just how far that distance is. Of course, people a hundred years ago would never travel that far in a day but now we do and it gives us a feel for the size of the planet so that makes it a length distance from which we can begin.

The next larger distance to think about is the distance between the Earth and the Moon. This is about thirty times the distance of the plane ride and so it would take equivalently some three weeks at the same airplane speed, or a few days in a NASA spacecraft. The distance to the Moon is thus imaginable: if you walked at four miles an hour non-stop without sleep it would take about eight years to arrive and another eight years to return. Nevertheless, the arrival on the Moon of astronauts Armstrong and Aldrin in July 1969 was one of the most memorable events of the last century. Only in part was it due to the distance to the Moon, it was equally due to the concept of humans walking for the first time on an astronomical object other than the Earth.

The Moon is visible in the night sky, and just as often present in the daytime thanks to its reflection of light from the Sun. The Sun is by far our nearest star and its radiated energy is crucial to the possibility of life on Earth. How far away is the Sun? It is about ten thousand times the length of the airplane ride and would take about twenty years to reach at the speed of an airplane. Not that any sane person would want to go there with a surface temperature well above that of molten iron. The Sun is about four hundred times further away than the Moon, and is already at such a large distance that it far exceeds anything with which we are familiar. This sets the scale of the Solar System with the Earth, rotating on its axis once a day, orbiting once a year around the Sun at a distance of some ninety-three million miles. Other planets like Mercury and Venus, circulate inside the Earth's orbit while six others including Mars, Jupiter and Saturn orbit outside the Earth.

It is almost inconceivable that any human being will travel outside of the Solar System in our lifetimes because of its enormous size. Yet on the scale of the visible universe, the Solar System is, in contrast, unimaginably tiny and insignificant. If there were no life other than on the Earth the universe would seem to be an absurdly large object if life were its primary goal.

In addition to the Moon and some planets, we can see thousands of stars with the naked eye. Most of these stars are similar to our Sun but appear much dimmer because of their distance from us. How distant are even the nearest stars? The answer is some two million times the distance to the Sun. So whereas we can reach the Sun in twenty years at the speed of an airplane, to reach the nearest star in twenty years would require a two million times faster speed. A quick calculation shows that this takes six hundred miles per hour into thirty-five thousand miles per second. To put such a speed into perspective, the speed of light is about one hundred and eighty thousand miles per second. This means our imaginary airplane, suitably coverted as a spacecraft, must travel at one fifth of the speed of light just to reach the nearest star in twenty years.

Here we see the limitations to any travel possibilities not only in our lifetime but what would seem to be forever. According to the theory of relativity, which there is no reason to doubt, nothing can travel faster than the speed of light. So even if the human lifetime is extended by medical advances to two hundred years or even a thousand years, it is impossible to travel during one lifetime to more than a few hundred times the distance to the nearest star. But the galaxy to which our Solar System belongs extends about ten thousand times the distance to the nearest star. So it would seem impossible ever to leave our particular galaxy — known as the Milky Way from its appearance spreading across the night sky.

There are however a couple of holes in this argument. First, according to relativity, time slows down as one travels when approaching the speed of light. Second, it is conceivable that some cryogenic method might be devised to slow down the speed of life and greatly enhance the effective human lifetime. Even so, to travel outside our galaxy does seem forever impossible and cosmology may remain just a spectator sport.

One hundred years ago it was generally believed that the the universe was comprised of only the Milky Way. The size of our galaxy is only ten thousand times the distance to the nearest star and already that is two hundred thousand times the distance to the Sun. Therefore, the galaxy size is two billion times (one billion is a thousand million) the Earth-Sun distance. The size of the galaxy seems to be relatively independent of time and so in ignorance of a universe very much bigger than a single galaxy, it was believed before the 1920s that the universe itself was static, neither expanding nor contracting.

When the General Theory of Relativity was proposed in 1915, this state of the observational knowledge stymied what could have been predicted, namely, the overall expansion of the universe. This expansion, which is a key feature of the universe and will lead us to the conclusion that it had a definite beginning, became an option only by observations somewhat later during the 1920s.

Now we arrive at the final leap in the distance scale. The visible universe turns out to be about four hundred thousand times the size of the Milky Way, very much larger than previously imagined. That is, not only is the Solar System of neglible size with respect to the universe but so is the entire Milky Way. In fact, in theoretical cosmology galaxies are treated as point particles. And the human race may be confined forever to be within one of these points!

We have seen that the size of the galaxy is tremendously larger, by a factor of billions, than the distance to the Sun. And then the visible universe is yet again so much larger than a galaxy that to study it each galaxy may be regarded as just a single dot within it. This should communicate in so many words an idea of just how big the visible universe is. Now we will show how we know that the present expansion (and possibly time itself) had a beginning some fourteen billion years ago.

ॐ ॐ ॐ ॐ

As has already been discussed, the size of the Milky Way has not expanded by even one order of magnitude since it was formed some ten billion years ago. Within the Milky Way the Sun and the Solar System appeared about five billion years ago. The Earth is a little younger, about four and a half billion years old. The point is the general arrangement of the Sun and planets in the Solar System has not markedly changed in the last few billion years. During that time we may regard the galaxy and its contents being of a constant size.

A truly astonishing revelation comes when we study the same question for the entire universe, including hundreds of thousands of the billions of galaxies outside of the Milky Way. The issue is: what is their typical motion relative to our galaxy?

Here it is important to understand a phenomenon well known in physics called the Doppler effect. It is a more familiar phenomenon for sound waves than for light. When a train blows its whistle and passes a listener the pitch of the whistle falls from a higher note to a lower note. In fact, not only the whistle but the entire train's noise exhibits the same Doppler effect. Why does this happen? It is because the motion of the train towards the listener compresses the sound waves to become a shorter wavelength and, because the velocity of sound is unaltered, to a higher frequency. Similarly, when the train is moving away the sound waves are stretched and the frequency lowers. The pitch for a stationary train would lie between the two pitches of the train approaching and receding. The shift in frequency is calculable simply in terms of the ratio of the speed of the train and the speed of sound.

ॐ ॐ ॐ ॐ

Exactly the same Doppler effect occurs for light. If a galaxy is approaching our galaxy, its light appears with a higher frequency. For the visible spectrum the highest frequency is for blue light so we may say that the light is blue-shifted. On the other hand, if the galaxy is receding from ours its light appears shifted to a lower frequency and

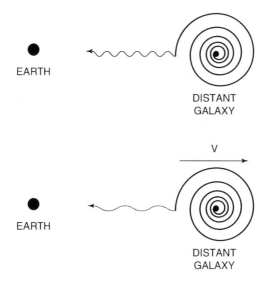

DOPPLER EFFECT: Wavelength of light from distant galaxy lengthened (red shifted) by speed of recessan from Earth

is red-shifted toward the red or lowest frequency end of the visible spectrum.

In fact, what are observed are the spectral lines of light emitted from known atoms and whose frequencies are accurately known here on Earth. If all the lines are systematically shifted towards the blue then the galaxy hosting these atoms is approaching the Milky Way: if toward the red then it is receding from us. This can be made precise by a mathematical expression for the frequency shift which gives the approach or recession speed as a fraction of the speed of light.

When this is studied for a large number of galaxies it might be expected that roughly half would be blue-shifted and half red-shifted if the overall universe were static and the galaxies were moving randomly.

What was observed, however, to the astonishment of Hubble and Einstein in 1929 is that almost all galaxies are red-shifted. Apart from a few galaxies in our immediate neighborhood like the nearby Andromeda galaxy all the hundreds of thousands of more distant galaxies measured are receding from us. This means that the entire universe is expanding and the galaxies are moving away from us and, as we shall see later, from each other. This phenomenon is called the Hubble expansion.

The next important question is: how does the recession velocity depend on the distance of the galaxy from the Milky Way? The galaxies can be classified into types such as spiral, elliptical or irregular and the total light emitted may be assumed to show regularity within each type. But their apparent brightness on Earth depends on the distance and falls off as an inverse square law. So the apparent brightness can be converted into a distance. When the recession velocity is compared to the distance a very important regularity appears. This is a most significant discovery in cosmology. It is called Hubble's law, which states that the recession velocity is proportional to the distance.

The ratio of the velocity and the distance is thus a constant, i.e., the Hubble parameter. Its value is notoriously difficult to measure but now we do know it to be close to seventy, within ten percent, in certain units. These units, which are not crucial to the general discussion, are kilometers per second per megaparsec, where a megaparsec is the distance light travels in about three million years.

How does this tell us when the present expansion phase began? This requires the use of the equations of the General Theory of Relativity together with two assumptions. The first assumption is that the universe at the large scale is the same, on average, in each of the three directions of space. This is called isotropy and is supported by observations which indicate that no preferred direction exists in the universe. The second assumption is that, on average, all positions in the universe are equivalent. This means that in all galaxies (hypothetical!) observers would see the recession of other galaxies according to Hubble's law. This assumption is called homogeneity. The combination of these two assumptions, isotropy and homogeneity, is technically known as the cosmological principle.

Combining the cosmological principle with the general theory of relativity gives rise to a mathematical equation known after its inventor as the Friedmann equation which characterizes the expansion of the universe in terms of a scale factor which is a function of time and specifies the typical distance between galaxies. Inserting the known Hubble parameter and the present known composition of the universe then enables us to calculate the scale for all past times. Note, in passing, that we cannot do this for future times with confidence because we do not know with certainty how the composition of the universe will evolve in the future. For the past we have good confidence and we find a striking conclusion: run in reverse, the contracting universe is seen mathematically to shrink to a point at a well-defined past time.

At that time the universe may have begun in some unimaginably powerful explosion called the Big Bang. Our present expansion seems to have a beginning and we know when. It was some 13.7 billion years ago, give or take two hundred million years. The age of the universe is now established to an accuracy of about two percent.

ॐ ॐ ॐ ॐ

There is one serious problem with extrapolating the Friedmann equation back in time, namely that about 13.7 billion years ago the equation becomes singular. The density and temperature become infinite and the classical theory of general relativity breaks down. Thus the Big Bang, only much more recently named, was the *initial singularity* to the early workers of the 1920s and 1930s.

Within the theory, we know therefore that the Big Bang must be avoided. A common response is to invoke quantum mechanics. From the fundamental constants, the speed of light, Newton's gravitational constant and Planck's constant one can construct a time known as the Planck time which is a tiny fraction (10^{-44}) of a second. One may say that at that short time after the would-be Big Bang the classical theory of general relativity must break down. Quantum gravity must play a role but till today, no completely satisfactory theory exists. So one may argue that quantum mechanics rescues the day. Indeed an entire field known as quantum cosmology has been built up around such an idea.

In this book we shall not appeal to quantum mechanics in this way but examine whether the Big Bang can be avoided in a purely classical context.

ॐ ॐ ॐ ॐ

Before proceeding there is one amusing anecdote about the origin of the graphic term Big Bang which seems apt to describe the beginning of the universe. Before the scenario we have just described was firmly established, a competing theory was the steady-state theory which postulated that despite the Hubble expansion there was a steady state and no beginning because of the continuous spontaneous creation of new galaxies. As a derogatory term for the competing theory, Big Bang was coined by a leading exponent of the steady-state theory. Unfortunately for Hoyle, it was the Big Bang theory and not his viewpoint which was confirmed by subsequent measurements.

ॐ ॐ ॐ ॐ

There is one alternative view (we shall discuss another view in Chapter 7) of the Big Bang where the lifetime of the universe is also infinite. The process of the Big Bang is in that view something which has occurred repeatedly, indeed infinitely, resulting in an infinite number of different universes of which we are aware of just one. This is technically called eternal inflation and the resultant universe becomes a multiverse for obvious reasons.

It is difficult, if not impossible, to test eternal inflation because the other universes would seem to be forever hidden from our view. Our best chance may be to make a probabilistic treatment of the multiverse to estimate the probability of the universe we observe having the properties it has in terms of its fundamental consituents or building blocks. Some research is indeed being pursued along this line.

ॐ ॐ ॐ ॐ

In this book we shall assume the beginning of the present expansion era to have taken place approximately fourteen billion years ago and followed, as will be discussed, by an inflationary era of rapid expansion. The latter explains two different kinds of extraordinary smoothness observed. We know that there was temperature uniformity to one part in one hundred thousand in the universe when it was only four thousand years old. Then there is the proximity of the observed density of the universe to a special value known as the critical density which would, without inflation, require preternatural fine-tuning in the early universe.

Inflation appears now to be ubiquitous in almost all theoretical cosmology, in one form or another. As we shall discuss, it can account for the otherwise puzzling smoothness properties of the universe. On the other hand, it is exceedingly difficult to make direct measurements which are sensitive to such an early era, the inflationary era, which occurred even earlier than a billionth of a second after the would-be Big Bang.

Normal observations involving electromagnetic radiation go back only to a few hundred thousand years after the would-be Big Bang, far too recent for us to study inflation. Studies of abundances of light elements like helium and hydrogen probe indirectly back to a cosmic time of one second after the would-be Big Bang. Potential neutrino astronomy measurements could directly probe a similar era.

The only chance of direct observation of the inflationary era would appear to be by gravitational radiation — waves created by strong gravitational fields in the early universe. The observability of such radiation depends on how early inflation took place, the earlier being the easier to detect. For later inflation it looks presently impossible to detect such gravity waves. The word "presently" is essential since how technology will evolve, and what consequent scientific apparatus will be enabled by the end of the 21st century is impossible to predict. It is a lesson from the history of physics that to decree anything to be impossible is a dangerous prediction.

ॐ ॐ ॐ ॐ

One may ask what happened before the Big Bang, if it did occur? This is beyond scientific investigation and it is easier to assume that time began then. Very ambitious and speculative theories discuss prior times using ideas such as T-duality in string theory or eternal inflation with its resultant multiverse. If such theories become testable and shed light on the physics of our universe, then they must be taken very seriously in a more general domain of applicability. At present, such ideas remain speculation.

Another question which we shall address at length in this book is what will happen to the universe in the future? This is less well-understood than the past, and depends critically on the properties of the newly-discovered Dark Energy which comprises almost three quarters of the total energy density of the universe.

Concerning space one will ask whether it too, like past time, will be finite in extent. Is it possible that by proceeding in a straight line one will return, after a finite time and distance, to the starting point due to a non-trivial topology of space? There is no compelling evidence for this possibility although certain data on the cosmic microwave background radiation can be interpreted as supporting such an assumption. Alternative explanations for the data come from arguments about cosmic variance or from small distortions in the hypothetical inflation potential, so the case for non-trivial spatial topology is not strong. If there were non-trivial topology, it could be of one of three types. Positive curvature corresponds to a closed universe, negative curvature to an open universe and a flat geometry, without curvature, as predicted by inflation. The local properties in such a universe satisfy the same general relativity equations as for the case of infinite space with trivial topology; only the global toplogical properties differ so it is not obvious from, say, the study of our galaxy alone which option Nature chooses. The notion of non-trivial topology of space necessarily introduces at least one fundamental length which, to be consistent with observational data, must be comparable to the radius of the visible horizon of a few gigaparsecs.

A common question by an educated non-physicist is: into what does the universe expand or, equivalently, what is "outside" the universe? So let us try to give a clear answer. The answer is not obvious only because of the limitations to the human imagination. All of us can easily imagine three spatial dimensions but four is exceedingly difficult. Unfortunately, the spacetime manifold of the universe is itself four dimensional and this is both why the question naturally arises and why the answer is slightly elusive. If we scale down by one dimension there is an analogous situation which is, by contrast, very easy to grasp. Take a balloon with spots on the surface to represent galaxies. As time passes we inflate the balloon and the spots get further apart as for the expansion of the universe. Now a two-dimensional being on the balloon surface may ask: into what is this two dimensional space expanding? The answer is that there is nothing "outside" the two-dimensional surface as obviously it is a closed surface without a boundary. Similarly the three-dimensional space of our universe has no boundary and no "outside".

࿎ ࿎ ࿎ ࿎

Among so many interesting yet unanswered questions, the one about the beginning of the present expansion 13.7 billion years ago seems settled. The expansion itself was universally accepted only in 1965 as a result of the discovery of the remnant background microwave radiation. The uncertainty of the future scenario for the universe is under much study as a result of the discovery of dark energy dating from 1998. Thus we are at a very exciting time in the subject.

The establishment of a finite time of the present expansion of about fourteen billion years is clearly of fundamental importance which could be equated by establishment of a finite spatial extent to the universe. There is absolutely no compelling evidence for such an idea although some data from the WMAP analysis of the cosmic background radiation, particularly the unexpectedly small values of the low multipoles, have been interpreted as suggestive of finite size and non-trivial topology.

Certainly these cosmological discoveries change our picture of our own history.

Finally, in our discussion of the universe's longevity, it is important that we use a linear time, rather than logarithmic time, in the above discussion. The two are dramatically different. Firstly, in logarithmic time the age becomes infinite. But the difference can be better seen in a concrete analogy.

Suppose we condense the entire cosmic history of fourteen billion years into one day of twenty-four hours starting and ending at midnight. First we use linear time. The nucleosynthesis takes place just a trillionth of a second after midnight; recombination and the surface of last scatter are three seconds later; galaxy formation starts around 1.40 am; the Sun is created about 4 pm in the afternoon and Julius Caesar invades Gaul about a hundredth of a second before midnight.

But if we map the same history using logarithmic time starting at the Planck time (since we must now start at a finite time in the past) then the occurrence of major events looks completely different.

Nucleosynthesis waits until 5 pm in the afternoon; recombination and the surface of last scatter are at 10 pm in the evening; galaxy formation begins at 11.25 pm; the Sun is created at 11.48 pm and Julius Caesar appears now only a trillionth of a second before midnight.

This illustrates how the use of linear time in cosmology effects the relative spacing of subsequent events. From the viewpoint of fundamental physics more happened in the first second of the Big Bang than has happened in the subsequent fourteen billion years, more in parallel with a logarithmic picture of time. But it is in linear time, with which we are familiar in measuring all everyday events, that the time since the would-be Big Bang does have the finite value of 13.7 billion years.

$$\approx \quad \sim \quad \infty \quad \ll$$

The answer to the question in this chapter's title has already been alluded to, that many other scientists (if not this author) explain away the initial singularity of the Friedmann equation by an appeal to quantum mechanics and quantum gravity. The absence of a fully satisfactory theory of quantum gravity can act as a further security for such scientists as no one can definitively refute the argument.

However, the tentative attempts at quantum gravity have problems. The concept of the wave function of the universe, as employed in quantum cosmology, is problematic for several reasons, not least of which is that the observer is inside the system. The Planck time is much shorter than the time expected to pass between the would-be Big Bang and the onset of inflation. Explaining away the initial singularity by quantum mechanical arguments was useful only when it was *faute de mieux*. This is no longer the case.

I find it more satisfactory to make a cosmological model where the density and temperature are never infinite. This precludes a Big Bang and replaces it with a different picture of time where time never begins and never ends. This is in contrast with the standard cosmological model where time begins at the Big Bang and never ends during an infinite future expansion.

While I cannot prove rigorously that the conventional wisdom is wrong, it does entail the singularities and concomitant breakdown of general relativity that we have mentioned. The existence of plausible alternatives now, however, makes the Big Bang idea less plausible.

As we shall show later in the book, there is an alternative version where time begins at a Big Bang and ends in a "Big Rip" at a finite time in the future. I regard this as preferable aesthetically to the conventional picture. But best of three possibilities about time is the "infinite in both directions, past and future" as exemplified by a cyclic model that my student and I constructed only in the twenty-first century based on the dark energy component. Observations of dark energy and its properties, especially its equation of state, will confirm or refute such more satisfactory ideas about time which insist there was never a Big Bang.

රⴚ CHAPTER 2 ❧

Smoothness of the Universe

When we look further at the universe at large, with the galaxies treated as point particles, there are even more surprises in store beyond those already discussed. Those discussed include the gigantic size, the cosmic expansion and the fact that the present expansion phase had a beginning.

What we have already discussed seems consistent, however, but not obvious to the naked eye. From looking at the night sky we might most naturally suppose, as people did for many hundreds of years before the twentieth century, that the stars had been there forever and would remain so. With the naked eye no further progress could ever have been made. Powerful telescopes see objects, or galaxies, far more distant and completely outside the Milky Way. They are moving away at high speed, higher as their distance increases. Ironically, the only other galaxy visible to the naked eye outside the Milky Way, the Andromeda galaxy, happens to be moving towards us! But almost all the other galaxies are receding from us.

To discuss the unexpectedly high smoothness of the universe, technically called the horizon problem, we need to introduce the concept of temperature. The universe is filled with radiation, electromagnetic radiation, which is currently extremely cold. It is at a temperature of about three degrees above absolute zero. The value of this temperature varies inversely as the scale factor which characterizes the size of the universe. Consequently, as the universe expands, the temperature is falling even lower. Turning that around, if we ask instead about the past, the temperature was higher.

In the history of the universe, at least as far as the electromagnetic radiation is concerned, a most important thing happened about three hundred thousand years after the would-be Big Bang at which time the visible universe was about one thousand times smaller than it is now. At that time the temperature was therefore one thousand times higher than now. This means it was very hot, at three thousand degrees. It turns out that this is the maximum temperature below which hydrogen atoms can survive as bound states. A hydrogen atom consists of a proton which forms the hydrogen nucleus around which one electron orbits. At temperatures above three thousand degrees the atom has a high propensity to ionize into a separate proton and electron. Indeed at all times before this special "recombination" time the protons and electrons existed separately in what is called an ionized plasma. After the special time the hydrogen atoms existed as bound states. Technically this special occurrence is illogically but forever called recombination. The name is illogical, it should be just combination, because the protons and electrons had never previously been combined!

Electromagnetic radiation is composed of massless elementary particles called photons travelling at the speed of light. Photons are scattered by charged particles but not by neutral ones. This is why the recombination occurrence is so important for the electromagnetic radiation in the universe. Before recombination there was an ionized plasma of electrons and protons. Such charged particles can scatter the photons. This means that the universe was opaque until recombination. Photons could not travel in straight lines at the speed of light because of scattering by charged particles in the plasma. After recombination, in contrast, the charged electrons and protons became bound into neutral hydrogen atoms. The universe became transparent as it allowed photons to propagate freely through it in straight lines since neutral atoms have no interactions with photons.

$$\text{ຂ} \quad \text{ల} \quad \text{ఛ} \quad \text{ઠ}$$

This means that the photons detected from the universe as a whole, technically called the cosmic microwave background radiation, have been able to travel along straight lines at the speed of light for the full fourteen billion years since the recombination occurred. This provides us with a unique and extremely valuable means of studying the state of the universe just three hundred thousand years after the beginning of the present expansion phase, or the would-be Big Bang. Much of the recent progress in cosmology is due to the vastly improved experimental information since 1992 on the details of this background radiation.

At a temperature of three degrees above absolute zero, the wavelengh of these relic photons is about ten centimeters or four inches. This is in the part of the electromagnetic spectrum called microwave and is of similar wavelength to the radiation used in a domestic microwave oven. When they started out at recombination, from what is technically known as the surface of last scatter, their wavelength was one thousand times smaller or a tenth of a millimeter. This is because the wavelength of a photon goes inversely as its energy. Energy is proportional to temperature and the temperature was at that time one thousand times higher than now. Another equivalent viewpoint is that the wavelength of the photon has been stretched by a factor of a thousand, like everything else, during its propagation to Earth from the surface of last scatter.

The background microwaves are invisible to the eye which can detect or see only a narrow part of the electromagnetic spectrum in the visible region where the wavelength is about a hundred thousand times smaller. But sensitive detectors of the microwave radiation have been flown on satellites and in the upper atmosphere suspended from large helium balloons. These detectors are looking out far beyond all the galaxies to the surface of last scatter which represents the most distant surface ever visible by electromagnetic radiation. To see further will require the study of either neutrinos or gravitational radiation.

To explain why there is something extremely remarkable about the observed smoothness of the universe at the surface of last scatter requires that we take a close look at relativity theory. One prediction of relativity theory is that nothing can move at a speed greater than that of light. This applies not only to particles but also to any type of information. In particular, a causative phenomenon may not create any effect faster than the speed of light. We are therefore influenced only by events in the earlier universe that are such as can transmit information to us. Technically, in special relativity theory we say that an event can be affected only by earlier events that are in its backward light cone. This requirement is called causality and is extremely restrictive on possible physical theories. Relativity theory has passed many tests and there is every reason to believe in it and in this consequent requirement of causality.

When the microwave radiation coming from the surface of last scatter is analysed it is found that its temperature distribution is astonishingly smooth. The temperature is exactly the same in all

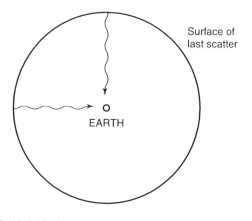

COSMIC BACKGROUND RADIATION: originated at recombination, about 400,000 years after beginning of present expansion at surface of last scatter.

directions to an accuracy of one in a hundred thousand. At that level there are exceedingly tiny fluctuations which are themselves extremely important in the formation of structure. But the important observation for the present discussion is that the distribution of temperature over this most distant surface is very, very smooth. Why is that so surprising?

It is surprising because in the simplest Big Bang picture it would violate the sacred principle of causality. Why is that? It is because the diameter of the present visible universe is roughly thirty billion light years and hence the diameter at the time that these photons were emitted was one thousand times smaller or about thirty million light years. A light year is the distance light travels in one year. The age of the universe at recombination was just three hundred thousand years so information could have traveled only three hundred thousand light years. The radius of the surface of last scatter is a hundred times bigger. So on the two-dimensional surface there are the square of one hundred, or ten thousand, regions of the surface which have never been in causal contact in the simplest Big Bang scenario. Therefore, it is amazingly enigmatic that the temperature distribution is uniform within all these ten thousand causally-disconnected regions to an accuracy of one part in a hundred thousand! This is the first example of smoothness.

It is very important to understand the previous paragraph because it suggests that the Big Bang theory has to be modified. Before describing the most popular modification of the Big Bang which accommodates this feature, we shall describe a second example of smoothness called flatness which will also be accommodated in the same modification to be described at the end of this chapter.

❧ ❧ ❧ ❧

The second example of smoothness observed for the present universe is different but has a similarity with the first in that it is a property of the present universe which, when we extrapolate back in time, seems utterly perplexing in the context of the Big Bang theory.

As already discussed in Chapter 1, the mathematical underpinning of the expanding universe is provided by the general theory of relativity combined with the assumptions of isotropy and homogeneity which make up the cosmological principle. This leads to a simple differential equation for the time dependence of the scale factor.

❧ ❧ ❧ ❧

This equation tells us that the ultimate fate of the universe would appear to depend on the overall density of energy and matter in the universe. Of course, we are talking only about the averaged density and not a local density where there is structure such as galaxies and stars. There is a critical value for this density which is such that the universe would expand forever but more and more slowly and slow to a stop after an infinitely long time. With this critical density another viewpoint is that the positive kinetic energy of the motions of the galaxies and other matter in the universe exactly balances the gravitational energy associated with the gravitational attraction. Energy is conserved so that the natural endpoint is where the galaxies come to rest at infinite separations and therefore there is no residual gravitational attraction.

If the density is less than this critical density, the universe's geometry has negative curvature and it will expand forever without coming to a halt. This negative curvature universe is technically termed open. The final possible case where the density is larger than the critical density has positive curvature. In it, the expansion will eventually halt followed by a recontraction to a Big Crunch. In other words, for such a closed universe the gravitational energy more than overcomes the kinetic energy. These simple arguments ignore the complications introduced by dark energy. We shall return to this issue.

The special case where the density equals the critical density is at the borderline between open and closed and is called flat. The three different geometries can be characterized by whether the angles of a triangle add to less than or more than, or equal to the canonical one hundred and eighty degrees. This is the total of the three angles in flat geometry, also called Euclidean geometry — the simplest form of geometry taught at school and very well known since the times of the Greek civilization. The other two cases are examples of non-Euclidean geometries, discovered by mathematicians in the 19th century. An example of positive curvature is provided by the surface of a sphere such as the Earth. Drawing a triangle with one vertex at the North pole and the other two vertices on the equator a quarter of a circle from each other gives a triangle that has all three angles equal to ninety degrees and a total of two hundred and seventy degrees. The larger total characterizes a positive curvature as in a closed universe. In a negative curvature as in an open universe, the three angles of a triangle sum to less than one hundred and eighty degrees.

The present total density of mass and energy is equal to the critical density with an accuracy of two percent. In other words, the angles in any triangle really add up to one hundred and eighty degrees to good accuracy. This fact is very surprising in the Big Bang theory and suggests that a component is missing. Why is that?

ॐ ॐ ॐ ॐ

In the equations which describe the evolution of the scale factor, the flat geometry is an unstable solution. This means that if the universe is flat now it most probably was always exactly flat. Deviations from flatness either in the open or closed direction become rapidly magnified as the universe expands. To be so close to flat now, the density one second after the Big Bang must have equalled the critical value to one part in a hundred trillion.

ॐ ॐ ॐ ॐ

We have described two different forms of extreme smoothness exhibited by our present universe in the large: the uniformity of the temperature over the surface of last scatter to an accuracy of one part in a hundred thousand despite the surface being comprised of ten thousand causally disconnected parts; and the fact that the geometry is flat after a long evolution of 13.7 billion years. *A priori* both of these circumstances are exceptionally unlikely. This latter unlikeliness can be compared to balancing a pencil on its sharpened point and finding it still so balanced a very long time later: any physicist or non-physicist would demand an explanation.

The appropriate modification of the Big Bang model is suggested by the study of theories for particle physics at high energies. Such theories involve dramatic transitions between different phases as the temperature falls in the early universe. As the universe cools it is possible for it to become temporarily trapped in a state with peculiar properties called the wrong vacuum. In the short time that it spends in this wrong vacuum or ground state it is normal that the universe undergoes a very rapid exponential expansion. To lead to adequate smoothness it is important that the rapid expansion be at least twenty-eight orders of magnitude in scale. This assumed period of super-fast expansion is called inflation.

Inflation offers an explanation of both of the smoothness properties, the horizon problem and the flatness problem, in one fell swoop. The horizon problem is solved because the entire present visible universe arises through inflation from just one tiny causally connected region. The flatness is explained because the inflation renders any pre-existing curvature negligible. This is like taking a balloon and inflating it (without bursting!) to an extraordinarily large size. The surface which at the beginning was strongly curved would become essentially flat at the end of the process.

Indeed the only way known to accommodate the two shortcomings associated with the smoothness properties of the Big Bang scenario is to assume an inflationary era in the very early universe. There is no direct evidence for inflation but the idea is widely regarded as likely to survive in some form or other and provides an important new ingredient in the Big Bang theory which became established in the 1960s.

One of the most intriguing aspects of inflation is that it provides a possible mechanism which disrupts the total smoothness initially by only exceptionally tiny perturbations. This is an effect of quantum mechanics during the inflationary era in the first billionth of a second after the would-be Big Bang. These quantum fluctuations exit from the comoving Hubble radius, characterizing the distance which is causally connected during the inflationary era, and can re-enter very much later after some hundreds of thousands of years or more. The question is whether these quantum fluctuations can lead to the later perturbations needed to seed the large scale structure of the universe. This requires our understanding, more clearly than at present, of the connection between the quantum fluctuations which exit the horizon and the resultant classical perturbations seen in the study of the cosmic background microwave radiation from the surface of last scatter. Nevertheless, it is a seductive idea that gigantic structures such as clusters of galaxies can be regarded as enormous amplifications of once ultra-microscopic quantum effects.

&ed; &ed; &ed; &ed;

Without inflation the extreme homogeneity and flatness of the present universe requires an extraordinary amount of fine-tuning of the initial conditions in the early universe. One may argue that since we are dealing with just one unique universe the starting point could be arbitrarily special. But physicists dislike fine-tuning since it reduces the power of the explanation. In this case, let us illustrate just how special the initial conditions would need to have been.

At the surface of last scatter corresponding to a redshift of about one thousand, we have seen that the density perturbations are at a level of one in a hundred thousand, already very tiny but accommodated in inflationary scenarios. Without inflation these perturbations would originate from very much smaller perturbations earlier in cosmic time. We may roughly expect perturbations to evolve linearly with the expansion of the universe.

Thus if we extrapolate back in time from the last scatter surface the density perturbations become progressively smaller. Going back to one second after the would-be Big Bang they have already diminished by a factor of a million so that they are then only one part in a hundred billion. This is so close to perfect homogeneity that it is an extremely fine-tuned situation, meaning that it is almost infinitely unlikely among all initial conditions. More generally, a theory is considered fine-tuned if there are extremely small (in this case the relative size of the perturbations) numbers which are not zero but whose size are not naturally explained. The technical term "naturalness" was invented for such a case. Naturalness implies the absence of fine-tuning of the parameters of the theory.

If we go back even further in time, the situation becomes all the more unnatural. At the Planck time the red-shift has increased by another twenty-two orders of magnitude so that the perturbations of the unadorned Big Bang model must then be at the level of a zero followed by thirty-two zeros and then a non-zero digit! Such an absurdly small number is taken as a sign that something is seriously inadequate with the theory. While it does not violate any physical law to assume such an initial condition, it violates common sense and is not what any self-respecting theorist will include as part of a theory. In other words, it will accommodate the observations but does not explain why there is such very high homegeneity in terms of any other physical principles.

❧ ❧ ❧ ❧

The situation concerning flatness is quite similar. As we have already discussed, the proximity of the present total energy density to the critical energy density implies that if we extrapolate back in time, the proximity becomes accurate to the extent that the ratio of the total density to the critical density becomes exactly one to an incredible accuracy.

Going back to one second after the Big Bang, this ratio must equal to one with an accuracy of one in a hundred trillion, and again this is an intolerable amount of fine-tuning from the theoretical physicist's viewpoint. At an earlier time it becomes even more intolerable because at the Planck time the proximity of the ratio to one must have been precise to better than one part in a trillion trillion trillion. This seems an inevitable part of the unadorned Big Bang picture. It does not refute that general picture, for which there is very much support including the nature of the cosmic expansion, the properties of the cosmic microwave background radiation and the successes of nucleosynthesis of the light elements.

What it does mean, however, is that an augmentation of Big Bang theory is needed, particularly for the very early universe, to replace the assumption of such extremely fine-tuned initial conditions.

ം⌒   ⌒ു   ൴   ⌒ം

One can imagine alternatives to inflation which could explain such fine-tuned initial conditions. One attempt has been to invoke the collisions of three-dimensional objects, or "branes", colliding together in a higher-dimensional space. This can have the effect of inducing the required amount of homogeneity and flatness just because the collision affects the whole three-dimensional space equally. So far such an "ekpyrotic" scenario has run into a wall of criticisms concerning the technical details although such an imaginative idea is worth exploring as a potentially viable alternative to inflation. Of course, the assumption of extra spatial dimensions is itself a big step to take! But only time will tell.

Yet another speculative possibility is to assume that the Big Bang originates from some pre-existing phase which transitions into the present universe in such a way that the necessary initial conditions are automatic. Such a "Pre-Big Bang" picture could shed light on the deep conceptual problems of the extremely early universe.

❧ ❧ ❧ ❧

Once we move forward in time from the surface of last scatter the density perturbations which are already present grow further and at the same time new ones enter the horizon. Gravitational instability leads to stronger growth of perturbations and the creation of large scale structures.

Structure is characterized by densities substantially larger than the mean cosmological density. At present the mean density is well measured and can be expressed in a variety of units. The dimension of density is of energy per unit volume and it turns out that it can be simply expressed as about ten electron volts per cubic millimeter. An electron volt is the energy acquired by an electron in traversing a potential difference of one volt. This mean energy density is of course extremely small by any everyday standard. It is one billion trillion trillion times smaller than the density of water.

As we look at a non-trivial structure the overdensity, which is the technical name for the mean density of the structure compared to the mean background cosmological density, increases. For a cluster of galaxies, the overdensity is about ten to one hundred while for a galaxy it is ten thousand. When we come down in size to the Solar System, the overdensity is closer to a hundred billion. To get an idea of how empty the universe is: if the Solar System were filled uniformly at the density of water it would have about the same mass as the whole universe! Bear in mind that the Solar System is infinitesimally small compared even to the Milky Way and completely negligible in size with respect to the entire visible Universe.

If we consider compact objects like the Sun or the Earth, the overdensity becomes the same huge factor of a million trillion trillion since their density is comparable to that of water, within a factor of a few.

The conclusion is: although there are such very strong fluctuations in density at cosmologically small scales, if we consider scales that are large compared to clusters of galaxies, say more than ten million parsecs, then the universe becomes to a good approximation exceptionally smooth both with regards to its homogeneity and it isotropy, being the same in all directions.

Such large scale smoothness is all important in concocting an appropriate theory of time and of the universe because it implies that the simple Friedmann equation derived from the assumptions of smoothness and of general relativity is very likely to be the correct choice, and we shall continue to make this assumption throughout the rest of this book.

CHAPTER 3

Structure in the Universe

Having raised the enigma of the smoothness of the universe on the large scale, it is now appropriate to address the opposite issue of why there exists structure in the form of very many galaxies, each containing very many stars. One such galaxy is the Milky Way containing our Sun around which our Earth orbits once per year. Recent work provides a plausible explanation for the origin and evolution of such structures. The structures are only a small perturbation of the smooth universe and yet are its most fascinating aspect. The explanation is based on the idea that the structures are "seeded" in the extremely early universe from effects of quantum mechanics, a theory formulated to describe the smallest scales associated with atoms and now relevant because of the enormous expansion of the universe since the Big Bang, to gigantic structures such as clusters of galaxies.

In the observations of the surface of last scatter there is a one in a hundred thousand deviation from complete smoothness. This tiny effect must lead to the structure formation. There are two aspects to explain: the primordial origin in the very early universe of these fluctuations and their subsequent evolution into the observed galaxies and stars.

We first discuss the origin of fluctuations and this requires a familiarity with quantum mechanics. The precursor of quantum mechanics is classical mechanics which was set out in the 1680s in the *Principia*, the book written by Isaac Newton. Based on simple laws such as force equals mass times acceleration, classical mechanics successfully describes the motions of objects resulting from pushes and pulls. This includes not only objects of everyday sizes, but when augmented by the law of classical gravity enunciated in the same book, objects like the Moon circulating around the Earth and the Earth orbiting the Sun. Classical mechanics was so successful that Newton's *Principia* dominated physics for the next two hundred years. It is impossible now to imagine any single publication dominating the physics community for so long.

In 1864, classical mechanics was extended further by Maxwell's theory of classical electrodynamics which successfully accounted for the motions of charged particles in electric and magnetic fields. The combination of classical mechanics and classical electrodynamics formed the foundations of theoretical physics as it existed at the end of the 19th century. This great edifice of knowledge was such a source of pride to physicists that at least one great physicist announced around the beginning of the twentieth century that physics was essentially over because everything was understood.

ॐ ॐ ॐ ॐ

Such hubris was short-lived. In the early part of the 20th century, two serious limitations to the applicability of the foundations of theoretical physics were found, both of which led to revolutionary advances.

The first limitation was associated with the concept of the aether. Aether was invented as a hypothetical medium through which electromagnetic waves could propagate. It seemed at the time inconceivable that such a wave could travel through vacuum and would need aether as an analog of the air through which sound waves propagate. But if there is such a thing as aether, the Earth is moving through it on its orbit around the Sun at a speed of about one ten-thousandth of the speed of light. Therefore, it was argued, the speed of light should be different whether the light travels parallel or anti-parallel or perpendicular to the Earth's motion. A precise interferometric experiment was completed in 1887 that intended to measure such differences. The surprising result was that the speed of light did not depend on the direction it was measured. The conclusion was

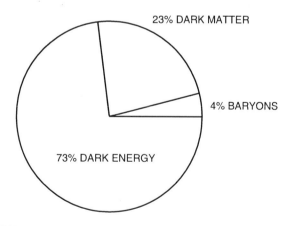

MAKEUP OF ENERGY CONTENT OF PRESENT UNIVERSE: 4% baryons; 23% dark matter; and 73% dark energy.

that the concept of some medium such as an aether at rest in the cosmos was erroneous.

This aether conundrum and the negative result from the interferometer experiment were in the air, so to speak, at the beginning of the 20th century and a crucial ingredient in the invention of the special relativity theory. Relativity leads to the prediction, already mentioned, that information cannot be transmitted faster than the speed of light as well as the prediction that the speed of light is always the same in vacuum for any direction, as well as for any source and observer.

The rise and fall of the aether idea may have a current analog. Many theoretical physicists believe that the understanding of dark energy, to be discussed in later chapters, could require another revolution as equally profound as relativity or quantum mechanics. Dark energy could play a role in theoretical physics in the 21st century of sparking a new revolutionary idea just as the aether conundrum did in the 20th century.

The second limitation of classical physics was the one which led to the invention of quantum mechanics in the 1920s. It was first noticed in the properties of heat radiated from a very hot body, and was heightened by a lack of understanding of the stability of atoms.

ॐ ॐ ॐ ॐ

For example, the hydrogen atom is comprised of a positively charged proton which acts as the hydrogen nucleus with a negatively charged electron orbiting around it. Regarding the much heavier proton as at rest and the electron circulating around it, then according to classical theory of electrodynamics, the electron must radiate continuously and lose energy. After a tiny fraction of a second it will spiral into the nucleus and the atom will collapse. The same is true for all atoms of the periodic table and so classical theory predicts that every atom in the universe can live for much less than one second. Needless to say, this total disaster for the classical theory in physics was not foreseen by the great man earlier announcing the end of physics.

This led, after some false steps, to the invention of quantum mechanics to replace classical mechanics especially for describing the inner workings of the atom. In the limit of large everyday sizes quantum theory reverts to the classical theory so that all the previous successes are maintained. But at the atomic size, one important prediction of quantum mechanics is that energy is not continuous but exists only in discrete amounts called quanta. Thus the atomic electron does not radiate energy continuously but in discrete amounts. Once the electron is in its lowest energy state no radiation is possible according to quantum mechanics and the stability of atoms follows. Quantum mechanics succeeded where the 19th century classical theory failed completely.

Quantum mechanics has many other successes but one particular feature merits our present attention. According to this theory there is an uncertainty in the value of all quantities which would classically be precisely knowable. For example, the position of the electron in the hydrogen atom cannot be specified like that of the Moon going around the Earth or of the Earth going around the Sun. Instead we know only a probability distribution, or wave function, which tells us what the probability of finding the electron in some region of space is. The total probability of finding the electron somewhere is necessarily unity just as in the classical case. It is just that on the atomic scales the position and speed of the electron have a quantum fuzziness which is a subtle aspect of the theory.

അ ഏ ഇ ഈ

A related phenomenon of quantum mechanics is that the vacuum is no longer empty but is alive with spontaneous creation and annihilation of particle-antiparticle pairs which may live a very short time, immediately to be replaced by others. To describe this situation requires a marriage of quantum mechanics and special, not general, relativity called quantum field theory. The essential point is that the vacuum value of a quantum field is continuously fluctuating rather than having a definite value as it would in a classical description. This uncertainty is very important for understanding the structure formation in the early universe. This concept has no counterpart in everyday life. For an intuitive picture in the imagination, think of a very sharply focused photograph as the classical theory and then put it very slightly out of focus. To capture the nature of quantum mechanics the out-of-focus scale needs to be of atomic size since everything is fluctuating by about that much and not more.

At this point, there seems to be no possible connection between quantum mechanical uncertainty and the stars and galaxies. But that would overlook the enormous amount by which the universe has expanded.

そ そ そ そ

In the inflationary scenario, as mentioned already, the vacuum is assumed to roll in a potential between a higher and a lower energy value as part of a phase transition between an unstable vacuum and a stable one. As the quantum field, technically called a scalar or more specifically an inflaton field, rolls down the potential, the universe undergoes a very rapid expansion. From a purely classical viewpoint, at the end of inflation the consequent surface of last scatter would have perfect smoothness properties with no variations at all to seed structure formation.

In quantum field theory, however, the field must fluctuate while rolling down the potential and these quantum fluctuations can lead to the tiny ripples on the surface of last scatter. There are some intermediate steps in realizing this idea but it may just work out. To be honest, the exact size of the ripples at one part in one hundred thousand cannot be predicted from inflation theory. That is the amount necessary to evolve into the structure observed much later as well as being the size of the fluctuations observed by COBE (1992) and WMAP (2003) satellite measurements of the cosmic microwave background radiation at the surface of last scatter. It is perfectly natural that such ripples at such a level originate from the quantum mechanical fuzziness during inflation.

The inflationary period is where the scale factor defining the size of the universe increases by at least twenty-eight orders of magnitude. If the expansion is exponential in time, it implies that the Hubble parameter is constant during inflation. The Hubble parameter is the ratio of the time derivative of the scale factor to the scale factor itself. A property of an exponential function is that its derivative equals to itself. The comoving visible horizon decreases as the inverse of the scale factor which means that the fluctuations go outside of the horizon during inflation. Another way of looking at it is that the fluctuations are expanded exponentially but the Hubble size remains constant. At a much later time, very long after inflation, the fluctuations re-enter the horizon. While outside the horizon the fluctuations are essentially frozen and re-enter as classical fluctuations ready to grow linearly with expansion and seed the structure of galaxies and the stars. On a clear night when one looks at the thousands of stars, and the Milky Way meandering across the sky, there is, only for the last twenty years, the new insight that they can originate from quantum fluctuations which happened 13.7 billion years ago.

These fluctuations from quantum uncertainty during inflation have three very specific and desirable properties. Firstly, they are adiabatic and seed density fluctuations systematically in various components like photons, neutrinos, baryons and dark matter which are present in the universe. Adiabatic is a term referring to the fact that no heat is transferred into or out of a comoving volume.

Secondly, the perturbations are Gaussian which means that the different wave numbers are all uncorrelated and each has a probabilistic distribution of a particularly simple and expected type. Thirdly and finally, provided that the potential governing the inflaton field during inflation satisfies conditions technically called slow-roll requirements, the spectral index which describes the relative power in different wave numbers is predicted to be close to one. All of these three properties are required to make a successful simulation of observed structure formation.

This is how inflation not only solves the smoothness properties, known as the horizon and flatness problems, but can even more impressively give a connection between quantum mechanics and the large scale structure observed in the universe. It is an appealing idea that the extremes of the very large and the very small can be so intimately related.

જે જે જે જે

Now we address the second question raised above concerning structure formation: how do the ripples on the surface of last scatter grow into the interesting and beautiful galaxies that we see?

The evolution of structure from the fluctuations as they enter the horizon involves nothing beyond Newton's law of gravitation as enunciated in the 1680s. This gravitational attraction acts principally on the nonbaryonic dark matter but also on the baryonic component some six times smaller.

From the time of recombination, when the visible universe was one thousand times smaller than today until when it was merely twenty times smaller, no significant structure of the baryonic part was formed. This era is picturesquely known to specialists as the dark ages. After recombination the photon and neutrino components remain negligible and evolve respectively into the current cosmic microwave background which is very well studied and a relic neutrino background which has yet to be detected. During the dark ages the fluctuations in all components begin by growing linearly with the expansion of the universe.

Eventually gravitational attraction causes local perturbations of nonbaryonic dark matter to grow nonlinearly and become more singular until, according to one scenario, the first baryonic stars can form with a mass of perhaps one hundred times the solar mass. Newton's law is adequate for this. Such massive stars collapse under their own gravitational attraction.

ॐ ॐ ॐ ॐ

As nonlinear effects dominate, the system must be analyzed semi-analytically or by computer simulation. Such simulations confirm that starting from the type of perturbations already discussed, the end results for the present universe using the known quantities of nonbaryonic dark matter and baryons can lead to a structure similar to that observed in galaxy surveys. The dark matter first clumps and then the baryons follow suit to yield clusters and superclusters of galaxies as well as the large voids which are observed. There are problems of detail at small scales; for example, simulations typically give too many small satellite galaxies for each large galaxy, and too much dark matter accumulates at galactic cores. But the three-dimensional distribution of structure on all larger scales gives rise to pictures which look to the eye, and more importantly to detailed statistical analysis, indistinguishable from the real universe.

To accommodate structure it is necessary to use both general relativity and quantum mechanics, though not together at any particular scale, with the assumptions of the cosmological principle and inflation. General relativity with the cosmological principle gives rise to the underlying geometry of the expanding universe. Quantum uncertainties in the inflaton field provide a possible and plausible primordial origin of fluctuations which exit the horizon in the very early expanding universe during inflation and re-enter the horizon much more recently. The evolution of these fluctuations into galaxies and stars requires only Newton's law of gravity which is not a separate assumption but a component of general relativity.

The theories of quantum mechanics and general relativity are applied separately to different regimes in this picture of theoretical cosmology. This is fortunate because there is no fully established marriage of general relativity and quantum mechanics into a consistent theory. The leading candidate is string theory and one very active area of research is the application of string theory to theoretical cosmology in a subject called string cosmology, a field so new that it is still evolving towards a more precise definition. Earlier attempts to connect string theory with particle phenomenology, which describes the interactions of quarks and leptons deduced at high-energy colliders, have so far been relatively unsuccessful. Whether support for string theory will come first from cosmology or from phenomenology is an interesting open question. Only time will tell.

ও ও ও ও

The standard model of particle phenomenology was first invented in the 1960s and 1970s. First a unification of electrodynamics and weak interactions was proposed by Glashow in 1960 and later completed in 1967 by Salam and Weinberg in a form that was conjectured by them to be a consistent quantum field theory, formally called renormalizable. The fact that the theory really was renormalizable was proven in the early 1970s by 't Hooft and Veltman. It was shown by them that the theory was equally as consistent and as amenable to precise unambiguous calculation as is quantum electrodynamics, the marriage of quantum mechanics with the classical electrodynamics theory of the 1860s.

In the 1970s, a parallel development was the evolution of a similar field theory of the strong interactions called quantum chromodynamics or QCD. The combination of QCD with electroweak theory comprises the standard model. Its detailed predictions have held up remarkably well. Even fourty years after the original proposal all experimental data, with one exception, agree with the predictions, up to an impressive one in a thousand accuracy. The exception is the non-zero neutrino mass first established in 1998 and which leads to the necessity of some modifications, still under intensive study, of the standard model.

అం అ అ అ

The standard model led to the successful prediction of new elementary particles including the charmed quark, and following the bottom quark discovery, of the top quark. There are six known flavors of strongly-interacting quarks which fall into the three doublets: (up, down), (charm, strange), (top, bottom). There are also three corresponding doublets of non-strongly-interacting elementary particles called leptons: electron, muon and tau with a partner neutrino for each type. These all group into three quark-lepton "families".

The standard model also led to the prediction of new types of weak interactions called neutral currents, discovered experimentally at CERN in 1973 and the weak intermediate bosons W and Z discovered, also at CERN, in 1983. All in all, this theory has been spectacularly successful and forms the basis of a theory for at least all the luminous matter seen in stars and galaxies and presumably all the baryonic matter. At the level of microscopic, subatomic scales it can be said that the standard model is a great achievement of theory in the second half of the twentieth century. It is therefore mandatory to ask how extending this successful model can accommodate features of observational cosmology, particularly the nonbaryonic dark matter.

৵ ৶ ৽ ৶

It has been a major industry for at least the last thirty years to extend the standard model in various directions. One idea is to unify strong with electroweak interactions into a grand unified theory. Studies of this type of theory at high temperatures actually led to the original idea for cosmological inflation. Just as with the uncertainty of how early, or equivalently, at how high a temperature inflation took place, there is an uncertainty in the energy scale at which grand unification happens. In the earliest and simplest such theories the unification scale was very high suggesting an extremely early inflation merely one trillion-trillion-trillionth of a second after the Big Bang. There are more recent unification schemes where unification as well as inflation could take place at much lower energies but, on whether such an alternative is correct, the jury is still out.

One other well-studied extension of the standard model is based on a serious technical issue within it. There is a scalar particle, the Higgs boson, necessary for accommodating the symmetry breaking between the electromagnetic and weak interactions, which gives rise to violent infinities called quadratic divergences. These seem to render the theory unnatural and lead to the consideration of an extension called supersymmetry. In such an extension each standard model particle has a new partner: a quark has a squark, a lepton has a slepton, and so on. It is not economical but does resolve the "naturalness" problem associated with quadratic divergences. It also leads to a candidate particle, known as the neutralino, which could constitute the nonbaryonic dark matter which is established to be making up almost one quarter of the cosmological energy density.

ॐ ॐ ॐ ॐ

The above outline merely gives a rudimentary idea of the very strong interrelationship between particle theory on the one hand and theoretical cosmology on the other. At an earlier time, say around 1980, particle theorists treated cosmology with some condescension because the cosmological data were so inaccurate compared to the reproducible precision data from high-energy accelerators. This has by now completely changed as the data on the cosmological microwave background radiation in 2003 has achieved high-enough accuracy to be characterized as "precision cosmology". The two studies of the very small (particles) and the very large (cosmology) have become inextricably intertwined. Most university physics departments now combine these two disciplines as a single group. The subject of string theory falls neatly into the rubric of "Particles, Strings and Cosmology," incidentally the title of one well-known series of international conferences. In academia, strings unify not only the aspects of particle phenomenology and theoretical cosmology but also stimulate common threads of research within physics and mathematics departments.

CHAPTER 4

Dark Matter and Dark Energy

In this chapter and the next we discuss the present make-up of the universe. Baryons are the stuff of which everyday objects are made so we start with a brief disussion of this component.

There are two methods to estimate the energy density due to baryons in the universe. One which was first used in the 1960s is by calculating the formation of helium and other light elements in the early universe about one minute after the would-be Big Bang. The other, which has been possible only since 2000, is by analysis of the relative heights of the odd and even acoustic peaks in the anisotropy of the cosmic microwave background. These two methods which therefore analyze quite different cosmological epochs agree very well with each other. The result is that the baryons make up about four percent of the total critical energy density. The visible luminous baryons corresponding to the stars that shine add up to less than one per cent so more than three percent is invisible and labeled baryonic dark matter. This four percent of the total energy density is the only part of which we have a clear understanding. The present understanding of the "dark" components should make theoretical cosmologists very humble but what an opportunity this presents for young people entering the field due to the fact that we have such limited understanding of ninety-six per cent of our universe.

ॐ ॐ ॐ ॐ

Nonbaryonic dark matter comprises about twenty-four percent of the critical density or some six times the baryonic density. Nonbaryonic dark matter is much more mysterious than baryonic matter because it is not something familiar in everyday life. Its presence has been strongly suspected already since the observations of Zwicky in 1933 but despite over seventy years of study, its nature is still a subject only of speculation. Nevertheless, two good candidates for nonbaryonic dark matter are undiscovered particles hypothesized by considering extensions of the standard model of particle physics.

ॐ ॐ ॐ ॐ

This standard model is a gauge field theory which describes the interactions of quarks and leptons. It has two pieces: one describing the strong interactions of quarks and known as quantum chromodynamics (QCD), the other describing a unification of electromagnetic and weak interactions known as electroweak theory. The two pieces are thus far unconnected. Nevertheless, it is seductive to attempt to unify the strong and electroweak pieces into a single theory technically called a grand unified theory. In such a theory all the three couplings for the strong, electromagnetic and weak forces unify at a very high energy far above familiar collider energies.

ও ৫ ৬ ৬

Such a grand unified theory has many attractive features and leads to remarkable predictions. The most striking prediction is the instability of the proton which implies that all objects are eventually unstable. Of course, the proton lifetime must be exceedingly long simply because material is very stable and even stronger limits are placed by the non-observation of proton decay in dedicated experiments. Nevertheless there are certain technical problems with the theory. For example, the electroweak theory contains a scalar field known as the Higgs boson which plays a crucial role in breaking the symmetry between electromagnetic and weak interactions. This scalar field must be included in the grand unified theory but then quantum corrections, technically called quadratic divergences, naturally force the Higgs field to have an extremely heavy mass near to the grand unified scale. But that destroys its capability to play the appropriate role in the symmetry breaking phenomenon. Another issue is that,

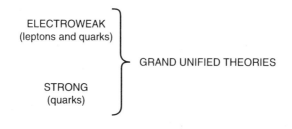

ELECTROWEAK
(leptons and quarks)

GRAND UNIFIED THEORIES

STRONG
(quarks)

GRAND UNIFIED THEORIES: Attempt to combine electroweak interactions of leptons and quarks with strong interactions of quarks.

phenomenologically, the unification of the three couplings does not occur very precisely within the minimal grand unified theory.

One way of ameliorating both of these two problems is to extend the standard model to what is called the supersymmetric standard model and to unify it in a supersymmetric grand unification. The additional assumption of supersymmetry is a mathematically elegant symmetry though for which there is no experimental evidence. It has the role of predicting a superpartner for each particle in the standard model: for example, for each quark there is a superpartner called a squark. The quark has spin one half while the squark has spin zero. Each superpartner has a spin differing by one half unit from its progenitor particle.

In the limit of exact supersymmetry, the particle and superparticle must have precisely the same mass. But this is experimentally excluded because, for example, a selectron at the same mass as the electron would be easy to detect and is clearly ruled out by experiment at that mass. Therefore supersymmetry is broken and the superpartners, if they exist, must in general be considerably heavier than their normal counterpart.

The assumption of supersymmetry has two advantages. First, when the Higgs boson is incorporated into a supersymmetric grand unified theory the Higgs mass is not affected by quadratic divergences. This is true even with supersymmetry breaking, provided such breaking is of a special kind known technically as "soft" and at a scale close to the scale characterizing electroweak breaking. Secondly, the unification of the three couplings becomes phenomenologically more precise when the superpartners are included than in the case without supersymmetry.

While admitting that the supersymmetry idea still has no support from experiment, there is no shortage of theoretical work along this line and the design of detectors for the next generation of colliders is arranged at least in part to find superpartners.

What has this got to do with nonbaryonic dark matter in the universe? This needs a little more explanation for us to understand how the supersymmetric standard model seems to provide a good candidate particle to play the role of nonbaryonic dark matter. In the supersymmetric standard model there is a natural discrete symmetry called technically R symmetry which means that the number of superparticles is conserved in every process. Such an R symmetry is necessary in the supersymmetric grand unification to avoid a proton decay lifetime too short for agreement with experiment.

One consequence of R symmetry is that there must exist a lightest supersymmetric particle (LSP) which is absolutely stable because there is no lighter superpartner into which it can decay while conserving the crucial R parity. This LSP is typically a linear combination of superpartners of the Higgs (Higgsino), the photon (photino), and the B gauge boson (bino). Such an LSP is generically called the neutralino.

The neutralino should have a mass of about one hundred times the proton mass and interact with strength characteristic of the weak interaction. As such, it provides an example of a class of particles invented for the purpose of comprising the nonbaryonic dark matter called Weakly Interacting Massive Particles (WIMPs). When one calculates how many such neutralinos survive annihilation in the early universe one is gratified to find that it fits perfectly with the amount of nonbaryonic dark matter which is observed. So the neutralino which follows from the assumption of supersymmetry is a good candidate to be all of the nonbaryonic dark matter. It works so well that some theorists regard this as a main motivation for belief in supersymmetry because the resultant neutralino gives such a natural candidate for a cosmological WIMP.

☙ ❧ ☙ ❧

There is a second candidate to play the role of the nonbaryonic dark matter arising from a different extension of the standard model of particle phenomenology, from a quite disparate line of argument. It is called the axion and its motivation is somewhat less than that for the neutralino but it is worth describing as yet another example of how there is such a strong interrelationship between theoretical cosmology and particle phenomenology.

In the theory of strong interactions or QCD, there is an issue concerning why the strong interactions respect certain discrete symmetries. These symmetries are parity (P) or mirror reflection and CP which is the product of parity with charge conjugation which interchanges matter with antimatter. Both of these symmetries, P and CP, are violated by the weak interaction but appear to be well respected in strong interactions.

Nevertheless there is a particular term which is allowable in the QCD theory which violates both P and CP and of which the coefficient must be of magnitude less than one ten-billionth to avoid conflict with experiments on bound states, particularly on the electric dipole moments of mercury atoms and neutrons.

One way to solve this problem is to impose and subsequently break an additional symmetry. This procedure gives rise to an additional particle called the axion. It can act as the nonbaryonic dark matter. The axion theory has certain technical problems such as that the theory appears inconsistent when combined with gravitational interactions.

The axion mass lies around a trillionth of the proton mass and so is a hundred trillion times lighter than the neutralino. These two candidates underline how little we understand the nonbaryonic dark matter: the range of possible mass for this missing ingredient actually lies somewhere between the axion mass and a million solar masses, a stunning range in mass of some sixty-nine orders of magnitude.

Thus, although the neutralino and axion are plausible candidates, the true solution chosen by Nature for the nonbaryonic dark matter remains enigmatic.

$$\approx \quad \approx \quad \approx \quad \approx$$

Given that almost one quarter of the cosmological energy density is nonbaryonic dark matter, roughly six times as much as the baryonic matter, the question is how to detect it? The arguments involving the study of galactic rotation curves, of the virial theorem in clusters, and of the general cosmological observations of the cosmic microwave background radiation, the large-scale structure and the type-1A supernovae are all indirect methods. They imply the fact that there must exist nonbaryonic dark matter. But of what is it comprised?

In particular, is it made up of weakly interacting massive elementary particles (WIMPs) with mass of order a hundred times the proton mass, or the much lighter axion, or is it made of far larger entities with the mass of the Sun or larger? It is a measure of our present ignorance that such extreme possibilities are still viable. Next, given that this nonbaryonic dark matter has a specific form, what are its interactions with the baryonic matter?

This issue of the interactions is especially relevant to the possibility of detection. The key question is how strong are the interactions with ordinary matter? We know that there is a gravitational attraction from the method of its indirect detection. If the WIMP were the neutralino state of supersymmetry then it must also have a weak interaction with ordinary matter. In that case one method of detection is to use bolometric detectors, typically large crytals cooled to very low temperature. If a WIMP particle strikes the crystal and interacts, it may deposit energy in the form of vibrations, or phonons, in the crystal. Such phonons could then be detected. Such searches have been made and even claims of a positive signal have occurred, but these claims have not yet been reproducible and are therefore generally disregarded. The detection of the nonbaryonic dark matter will be an extraordinary claim. Extraordinary claims need backing by extraordinary evidence which so far does not exist.

Another possibility for detection is from the process of annihilation of dark matter with dark antimatter. If we assume this produces normal photons and electron-positron pairs, then the subsequent positron annihilations could be detected by their gamma-ray emission. Indeed there is a suspiciously large number of positrons near to the center of the Milky Way and it has been suggested that these may originate from the annilihilation of dark matter. Unfortunately, there are other explanations for the occurrence of such positrons and so the evidence for direct detection of dark matter is not yet compelling from that source either.

The direct detection of nonbaryonic dark matter could be rendered extremely difficult (one must never say impossible!) if its interactions are only gravitational simply because of the weakness of the gravitational interaction. For example, in such a case even if dark matter does annihilate against its antimatter, it would produce only more dark matter in the form of "dark photons" or whatever dark particles are relevant. It is only an assumption that the nonbaryonic dark matter posseses significant non-gravitational interactions with ordinary matter.

In such a case we might have to be content with indirect detection for some time but that would be frustrating in the sense that the identification of the constituent particles of nonbaryonic dark matter would remain elusive and would not act to motivate the building of models extending the standard model of particle phenomenology.

A more optimistic scenario is that there are significant non-gravitational forces and that direct detection will occur in the near future. This would provide an important guide to model building especially when the nature of the interactions between the dark and ordinary matter are further explored experimentally.

Enthusiasts for supersymmetry are naturally encouraged by the fact that identification of the WIMP with a neutralino very naturally gives the correct dark matter density after the WIMP-annihilations have taken place. Indeed this is taken by some as a strong suggestion that supersymmetry is correct. Another such evidence is the improved unification of the couplings in a supersymmetric grand unified theory.

On the other hand, the main motivation for supersymmetry is to provide naturalness for the Higgs boson in the sense of canceling quadratic divergences in the underlying field theory. If such naturalness can be obtained without supersymmetry then for it motivation would be removed, although supersymmetry still plays an important role in the construction of superstrings as a potential theory for quantum gravity. In that case, however, the supersymmetry can be broken at a very high energy scale, perhaps as high as the Planck scale, and there would then be no reason to expect to see any sparticles at masses accessible to the next generation of colliders.

It can be seen from such arguments that the direct detection of dark matter in non-accelerator experiments could help clarify precisely what particles we can expect to be produced at high-energy colliders. It is in this sense that the interrelationship between cosmology and particle theory becomes even more important.

ॐ ॐ ॐ ॐ

We have so far discussed the baryonic component of four percent and the nonbaryonic dark matter component which is twenty-four percent. Yet the total energy is consistent with the critical density.

This means that some seventy-two percent of the total mass-energy density has not been listed.

This additional dominating component has been discovered only since 1998 and is called dark energy. What is the difference of dark energy from dark matter?

To understand the distinction, it is necessary to introduce the ideas of pressure and equation of state. Baryonic matter and dark matter exert zero pressure, the corresponding particles are essentially at rest and the equation of state which is defined as the pressure divided by the density is equal to zero. Photons which necessarily travel at the speed of light do exert a positive pressure and their equation of state is equal to plus one third. These two values of zero and plus one third are the most familiar equations of state applicable to dust and radiation respectively. Dark energy is strangely different, and in particular, exerts a negative pressure. In its simplest manifestation dark energy corresponds to a cosmological constant with equation of state equal to minus one, or pressure equal to minus the density.

Negative pressure has no example in everyday life. Normally, with positive pressure, exerting a force on the piston confining a cylinder of gas will compress the gas and increase the positive pressure. For a cylinder of dark energy, however, the force would increase the volume, completely contrary to the physical intuition we develop from everyday experience.

We shall discuss much more about the equation of state for dark energy in the final chapters of the book. Here we recall the early history of the cosmological constant, the precursor of modern dark energy.

After general relativity was first published by Einstein in 1915, he followed up with its application to cosmology in his very important paper of 1917 which may justifiably be regarded as the start of theoretical cosmology. At that time, it was generally believed that the Milky Way galaxy was identical to the entire visible universe and that this universe was a static situation without overall expansion or contraction. We now know that the universe is really four hundred thousand times bigger and that the expansion can be seen only in the much larger theater in which the Milky Way and all other galaxies are treated as mere points.

Ironically, an observational astronomer named Slipher had published a few years earlier, in 1912, his result of observing more red shifts than blue shifts. It seems likely that Slipher was seeing the expansion of the universe but he naturally thought he was looking at stars within the Milky Way rather than at other galaxies outside the Milky Way.

ॐ ॐ ॐ ॐ

In reality, Einstein in 1917 missed a golden opportunity to predict the cosmological expansion which follow quite straightforwardly from his general relativity theory combined with the cosmological principle. Instead, he added a negative cosmological term which made the theory less elegant but allowed a static universe. Only much later in 1929 did Einstein realize from Hubble's publication that the universe is really expanding. He must have, at least figuratively, kicked himself for not sticking to the most elegant version of general relativity.

A moral of this story is that a theorist, including Einstein, may not take his own theory seriously enough. A sequel is that since 1998 the effect known as dark energy which resembles a cosmological constant, though now with a positive sign, has appeared and looks more than likely to remain as a robust part of cosmological theory.

ॐ ॐ ॐ ॐ

Should one regard this new development as something which vindicates the mistake of Einstein in not leaving the term out from his cosmological application of general relativity? This new discovery of an effect looking like a cosmological constant should not be attributed to the far-sightedness of Einstein. It arises not from the expansion of the universe but from the even more surprising fact that the rate of expansion is accelerating. No one had any notion of this even twenty years ago, let alone in 1917, and it has no connection to Einstein's motivation for adding such a term in his theory.

Rather it shows that there is only a small finite number of ways to modify the theory of general relativity while preserving all of its symmetry principles. It is surprising that the most economical version fails but future understanding of the additional piece associated with dark energy may shed light on how to discuss gravity correctly and to relate it to the other interactions.

ے   ے   ے   ے

One ambitious attempt at a consistent theory of quantum gravity is string theory and the place where one would have hoped for insight from string theory is certainly in the issue of the cosmological term. This has not been forthcoming and string theory actually seems slightly to favor a negative rather than the observed positive sign for the cosmological constant. This is still probably not the end of the string theory input, mainly because the theory is still plagued by the present inability to select between an astronomical number of candidate lowest-energy states. It is unclear whether this hurdle to progress will be overcome in the near or distant future in order to confront string theory with the real world. In any case, even if string theory turns out not to be the correct theory of quantum gravity, it has already, by a remarkable technical argument going by the name of "correspondence" between string and gauge theories, provided very interesting possibilities for gauge theories that could address some of the shortcomings of the standard model of particle phenomenology.

The tiny cosmological constant and the cosmic coincidence of the dark energy density now being comparable to the matter density may be explained by field theory models or alternatively by exotic "phantom menace" models with an unexpected equation of state as well as bizarre consequences for the distant future fate of the universe, as we shall discuss in further detail especially in the final chapters.

Composition of the Universe's Energy

The newcomer to the cosmological mass-energy menu, and its dominant component, is the dark energy. Since dark energy is considerably more obscure even than dark matter, and because dark energy could for the 21st century play the role played by aether for the 20th, the way in which dark energy was discovered and quickly established deserves discussion as it is a fascinating story.

In 1929 Hubble discovered that the recession velocity of a galaxy is proportional to its distance. The constant of proportionality, the Hubble parameter, is now known within a few percent error to be seventy in the units mentioned earlier: kilometers per second per megaparsec.

What had been assumed until 1998 was that the rate of expansion was decelerating. With no cosmological term, whether the universe were open, flat or closed the natural expectation from general relativity is that the rate of expansion is slowing.

One of the major problems with the evaluation of the Hubble parameter is the reliable estimation of distance to other galaxies. Until relatively recently this required a sequence of steps called the cosmological distance ladder, each step introducing its own error. Finally, the resultant distance and the corresponding Hubble parameter could still be uncertain by a factor two.

This distance ladder can be avoided if there is a so-called standard candle with known luminosity at a very great distance, ideally from when the visible universe was, say, half its present size. Such an object is provided by a supernova of a particular type, Type 1A.

A supernova is a gigantic thermonuclear explosion which happens when a star collapses under its own gravitational attraction. The most straightforward case is called core-collapse and does not involve interaction with any other object. A particularly consistent type involves instead one member of a binary pair of stars when one of them becomes a supernova. This is called technically a supernova of Type 1A, because it has no hydrogen in its spectrum (hence not Type 2) and because it accretes mass from its binary companion (hence not Type 1B or 1C).

By the study of nearby Type 1A supernovae it is found that the peak luminosity is simply related to the rise and fall time in luminosity. This regularity means that we can estimate the absolute luminosity and hence distance, by the inverse square law, to such Type 1A supernovae. The red shift and hence the recession velocity can also be measured and so very distant Type 1A supernovae provide points on the Hubble plot which are exceptionally distant and, because they act as standard candles, bypass the error-inducing process of the cosmological distance ladder.

The key point is that supernova Type 1A can be discovered as its light output is increasing, then observed as it reaches a maximum and then declining gradually in intensity. The rise and fall in intensity typically takes from a few weeks to a few months. Since it is not known where and when a supernova will occur, it is important to make the discovery to have a wide-angle telescope which may be quite small, typically four meters in diameter. Once the discovery is made, a large land-based telescope up to ten meters in diameter or, in space, the Hubble Space telescope can follow the subsequent evolution with greater accuracy.

What has been established phenomenologically for the nearby supernovae Type 1A is that the duration of the light curve characterizing the increase and decrease of the light intensity is closely correlated with the absolute value of the peak intensity: the longer the time period, the brighter the supernova and vice versa. Thus, observation of the time dependence of the light intensity translates into the measurement of the peak intensity and hence the distance. From a purely theoretical perspective, the theory of supernovae Type 1A is not sufficiently developed to derive this relationship between light curve and peak intensity but empirical data strongly supports the correlation.

During the 1990s such Type 1A supernovae were pursued by two groups of observers and the data analyzed to give a consensus on the unexpected result. Instead of slowing down as everyone had expected, the expansion rate of the universe is accelerating. Further analysis suggested that this acceleration has been happening for only a fraction of the time since the Big Bang and that the transition from decelerated to accelerated expansion happened at some time more recent than when the visible universe was half its present size.

These data suggest that there is a significant component of the total energy density which exerts a negative pressure tending to blow the universe apart. One example is a positive cosmological constant.

Fortunately there are independent checks for such a dramatic and unexpected result. If it were only the supernovae, it could be that they appear dimmer than expected merely because of obscuration by intermediate dust clouds. Alternatively there could be a systematic evolution effect between the very distant and nearby supernovae. But both of these possibilities already seem unlikely. For example, the obscuration would be expected to affect different colors differently and there is no sign of that. Also the similarity of spectra between the distant and nearby Type 1A supernovae would argue against any significant evolutionary effect.

The observations of these Type 1A supernovae was undertaken beginning around 1992 especially by Perlmutter's group based in Berkeley. Presumably, it was not anticipated at that time the extent to which such a study would revolutionize theoretical cosmology. In 1995 a second group based in Harvard was converted to join the chase as an independent group to check on the nature of the output results. Despite some false starts, the two groups eventually agreed on the result that the expansion of the universe is accelerating. This result was completely unexpected and so it was very important to have the two independent groups making the same observations and analysis.

అ అ అ అ

A second independent approach is to analyze the anisotropy of the cosmic background radiation. By decomposition into multipoles, one can plot bumps, technically called acoustic peaks, which reflect the acoustic oscillations of the baryon-photon plasma immediately before recombination.

The detailed positions and heights of these acoustic peaks are sensitive to the values of the energy components and so can give independent evaluations for the amount of dark energy as well as of baryons and the dark matter. This is because the photon trajectories from the surface of last scatter to Earth depend on the geometry of the intervening spacetime, whether it is Euclidean or non-Euclidean and, if the latter, whether the curvature is positive or negative. It is possible to find analytical expressions for the acoustic peak positions in this way. In addition, there are public software codes which will calculate the detailed shapes of the peaks for any input cosmological parameters.

❧ ❧ ❧ ❧

Physically we may say that the causally connected horizon represents only a small fraction, on the order of forty-thousandth of the full visible horizon. The angle subtended at the Earth by that causally connected region depends on whether the photons traverse straight lines, as in a flat geometry, or if their trajectories are bent by a curvature of space. This will affect which multipole is enhanced in the observations of the anisotropy. For a flat geometry, the prediction which depends on the time of recombination is that the multipole equal to about two hundred will be the location of the first acoustic peak. Further harmonics will appear at multiples of this value in the first approximation.

If the curvature is positive, the peak will appear at a lower multipole while if curvature is negative, it will appear at a higher multipole so this measurement gives a direct estimate of the nature of the geometry back to the surface of last scatter.

❧ ❧ ❧ ❧

The baryon content effects the relative heights of the odd and even acoustic peaks in a calculable way. The most detailed survey was published in 2003 from the WMAP satellite. It confirms that seventy-two percent of the total energy is in the form of a dark energy, while four percent is in baryons and the remaining twenty-four percent in dark matter. The errors in these percentage estimates are quite small.

The WMAP data does not pin down the equation of state for the dark energy. Although it is consistent with the value of minus one as for a cosmological constant, it allows values both more positive than this and interestingly also more negative, down to -1.1.

$$\approx \approx \approx \approx$$

Finally, there is a third independent means of analysis. From the large scale galaxy surveys such as the so-called 2dF Galaxy Redshift Survey and the Sloan Digital Sky Survey, the total amount of dark matter plus baryons can be deduced. The result is some thirty percent of the critical density with an error of some three percent. This is in agreement with the other two estimates. We note that each of these two galaxy surveys included above 500,000 galaxies.

To understand better how the dark matter component is estimated, it is necessary to examine the gravitational attraction and behavior within a cluster of galaxies. There is a general theorem, applying to any such gravitationally bound system, technically known as the virial theorem, which relates the kinetic energy to the potential energy. This is used to estimate the amount of gravitational mass in a cluster. By comparing this with the estimated mass of the luminous matter, one can check for any excess and identify the dark matter.

At the length scales of cluster and superclusters, just as for individual galaxies, it is found that dark matter is present at a level some thirty times the density of the luminous matter, or very roughly thirty percent of the critical density with an error of three percent.

If we input the observation that the geometry is flat from the CMB analysis then there is clearly a missing component, other than the luminous and dark matter, to be identified with the dark energy.

The important point is that the study of galaxy redshifts and particularly the virial theorem for clusters of galaxies points to the necessary presence of dark matter. At larger distance scales it becomes more important, but the consensus is that it does not approach the critical density even for scales approaching the size of the visible universe. Instead the convergence is on an amount of dark plus luminous matter approximately thirty percent of the critical density. The precise WMAP data confirms a number twenty-eight percent for this quantity.

$$\approx \quad \approx \quad \approx \quad \approx$$

Taken as a whole there is a cosmic concordance which is difficult to dispute. All three approaches lead inevitably to the conclusion that there is almost 3/4 of the universe (seventy-two percent) in dark matter, twenty-four percent nonbaryonic dark matter and four percent baryons.

Only the four percent is something we really understand! The remaining dark side of the universe is truly enigmatic. Even the baryonic component has a majority which is not directly detectable. At most one percent of the critical density can be accounted for by the luminous matter in stars and galaxies. The remainder of the baryonic component must be in the form of objects too cold or light to sustain thermonuclear power. These could be Jupiter-sized objects, brown dwarfs, etc.

Nonbaryonic dark matter could be composed of objects from the LSP particle, a kind of WIMP, with a mass of about a hundred times the proton mass all the way up to objects with a million times the solar mass. This range of masses which stretches over sixty orders of magnitude reveals our ignorance on nonbaryonic dark matter.

Even this level of ignorance pales by the side of the excitement created by the discovery of dark energy. This could be merely a reappearance of Einstein's cosmological constant, albeit with the sign reversed, and with a totally different motivation from Einstein's. If so, there is still a fundamental issue of fine-tuning which remains unresolved.

If the dark energy is something more exotic, then its understanding may well require a revolution in theory. That is why the dark energy provides so much excitement. The fact that it is so badly understood provides motivation for pushing again the limits of human intellectual creativity.

The end result of the cosmic concordance is that the make-up of the menu of the total cosmic energy density is therefore 4% baryonic matter, 24% nonbaryonic dark matter and 72% dark energy. The errors on these partitionings are quite small and it now seems well established from the cosmic concordance that this picture is accurate and unlikely to change significantly in the future.

The surprising and even slightly embarrassing aspect of this over-all composition, and one should add enigmatic, is that only the 4% contribution of baryons is well understood. The remaining 96% of the dark side of the universe is amazingly little understood. It stands as a wonderful opportunity for the upcoming generation of physicists to seize upon and make their reputation on understanding dark matter and energy.

The dark matter may be already implicit in certain extensions of the standard model of particle phenomenology, for example, the lightest supersymmetric partner in the supersymmetric extension of the standard model. For the dark energy on the other hand, it may require some further understanding of the fundamental theoretical physics to make real progress in its description.

෩ ෨ ෯ ෬

From observations in the near future it will be possible to pin down the equation of state for dark energy. This means the ratio of its pressure to its density. The density is always positive but for dark energy the pressure is significantly negative and the question is: how negative? It is known that the equation of state must be less than negative one third to accommodate an accelerating cosmic expansion. It is important to know whether the equation of state is greater than, equal to, or more negative than negative one. The future of the universe depends strongly on this equation of state of the dark energy. Its evaluation will require more accurate data on the cosmic microwave background radiation and on Type 1A supernovae.

We shall discuss the future fate of dark energy and thence of the universe in the next chapter but let us briefly preview here one of the more fascinating aspects for the case where the equation of state lies below minus one. In this case there is the peculiar feature that the universe ends at a finite future time when the scale factor diverges to infinity. This has been termed in the physics literature as the Big Rip. The dark energy density increases without bound as a positive power of the scale factor, also becoming infinite at the time of the Big Rip.

In a philosophical sense, this picture is quite attractive because it introduces a kind of symmetry between the past and the future in the sense that both are finite in linear time.

The physics of the final billion years before the Big Rip is quite interesting even if it is extremely unlikely that, if it occurs, anyone will be around to observe it. It has been computed how the dark energy density increases to such a level that, assuming that there are non-negligible fluctuations in it, it will gradually rip apart the existing structure.

A few billion years before the Big Rip, clusters of galaxies are the first to be erased. Galaxies are similarly dismembered some sixty million years before the Big Rip. Three months or so before the end, the Solar System becomes unbound. Just a half hour before the Rip the Earth will explode. Finally, a ten-million-trillionth of a second before the final gasp, all atoms will be torn apart and dissociated.

If this scenario is correct, we are living in a cosmological era some time between the formation of intricate structure from an exquisitely-smooth universe and a future time when all the structure will be torn apart just before a Big Rip. So it is no surprise that we are living when there exists the most complex structure.

೨ ೪ ೕ ೖ

All the above was concerning the dark energy properties and we discussed dark matter in the previous chapter. On both dark aspects our knowledge is inadequate so perhaps we should look again at the small four percent of baryonic matter which we do understand considerably better, and therefore there is much more that we can say.

The baryonic matter, like everything else, experiences gravitational interaction which is classically well described by the general theory of relativity. The non-gravitational interactions are well described by the standard model of particle phenomenology.

But even this standard model has plenty of open questions, not the least of which is the large number of free parameters which must be matched to experiment. With massless neutrinos there were already nineteen and now, with the massive neutrino sector, that number of parameters has increased to twenty-eight. Let us go through these parameters which some all-enveloping theory would be expected to explain.

॰੭ ੭੭ ੭੭ ੭੭

The masses of the quarks and leptons are twelve quantities which remain mysterious. Apart from the apparent ordering into three quark-lepton families, they might as well be twelve random numbers. Perhaps they are, but physicists would much prefer an underlying theory to predict them. That would be a very major step in understanding.

There are mixing angles and phases: three angles and a phase for the quarks; three angles and three phases for the leptons. These ten bring the total to twenty-two. The remaining parameters include three gauge couplings: one for QCD and two for the electroweak theory. Finally there are the Higgs boson mass, the weak scale (or the mass of the W boson) and, the most peculiar of all, a parameter called theta bar which controls an unwanted symmetry violation in the QCD of strong interactions.

Even if all these twenty-eight parameters could somehow be derived the standard model still remains inadequately unified because the quite similar theories of QCD and electroweak forces are disjoint and it is tempting to put them together in a so-called grand unified theory. One striking test of this idea is the prediction that the proton, and hence all matter, is unstable. The lifetime is predicted to be at least a billion trillion times the age of the universe. An experiment cannot wait that long, of course, but instead one can observe the astronomical number of protons in thousands of tons of purified water for a few years with similar sensitivity. Such water is usually surrounded by large photomultiplier tubes and placed deep underground. If a proton decays, for example into a pion and a positron, the decay particles

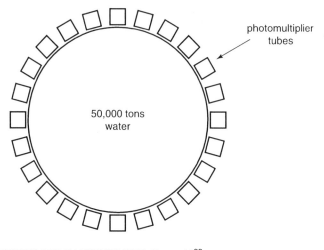

PROTON DECAY EXPERIMENT: Some 10^{33} protons
in water studied for a year show proton lifetime of at
least 10^{33} years.

will travel faster than the speed of light in water and consequently produce radiated light to be detected by the tubes.

Such experiments have so far failed to detect proton decay, although the extraordinarily sensitive water detectors have made revolutionary discoveries concerning neutrinos. For example, the first extrasolar neutrinos were detected in this way in 1987 from a supernova in the Large Magellanic Cloud lying outside the Milky Way. Similarly, it was first demonstrated compellingly in 1998 that neutrinos have non-zero mass by the study of the angular distribution of events produced by cosmic rays in such a giant underground water detector in Japan originally designed to look for proton decay.

The next step in developing the standard model is likely to be prompted by the data generated in the Large Hadron Collider at CERN, scheduled to start producing data in the year 2010. It is expected that entirely new particles will be discovered, including the Higgs boson and hopefully others. The others could be those predicted by supersymmetry or large extra dimensions. Perhaps they will be surprising additional states such as gauge bosons, fermions and scalars somehow arranged to cure the unnatural quadratic divergences in the scalar sector of the standard model.

❧ CHAPTER 6 ❦

Possible Futures
of the Universe

It might be hoped that with such a detailed understanding of the history of the universe we are in a position to predict confidently its future. We shall discuss in this chapter the fact that this is not at all the case.

Such a prediction would be possible were the dark energy component absent so let us first discuss that hypothetical case. The Friedmann equations that arise from general relativity under the assumption of the cosmological principle assign a special role to the critical density. If the total density of baryons and nonbaryonic dark matter are less than this, then the universe is open with negative curvature and will expand forever without coming to rest. The rate of expansion will decelerate.

If the total baryons plus nonbaryonic dark matter density is greater than the critical density, then the universe has positive curvature and will eventually stop expanding after a finite time and then recontract to a Big Crunch. Finally, if the baryonic plus dark nonbaryonic densities add to exactly the critical density then the universe is flat, the kinetic energy equals the gravitational potential energy and so it will expand forever coming to rest at an asymptotic infinite time. Notice that the division between baryons and nonbaryonic dark matter is irrelevant in all three scenarios since they all have the identical equation of state, namely zero pressure. In all cases, the rate of expansion will decelerate.

ॐ ॐ ॐ ॐ

The presence of dark energy spoils such simple prognostications mainly because the equation of state for dark energy is unknown. Not only is it unknown but it may depend on time, giving even more possibilities for the future.

The simplest possibility is that the dark energy equation of state is constant and equal to minus one. This is precisely the case for a cosmological constant.

For this possibility, while the dark matter density falls according to the inverse cube of the scale factor, the dark energy density remains constant. Therefore clearly the dark energy becomes more and more important as time progresses. At present, by a coincidence, the dark energy and dark matter have approximately equal roles. In the distant past the dark matter dominated while in the distant future dark energy will take over.

In this case, the scale factor will grow exponentially with a constant Hubble parameter and no recontraction can be foreseen. The universe becomes more and more dilute in its energy density. Thus, although dark energy played little role in the formation of structure in the past, it certainly plays the major role in deciding the future fate of the universe.

This is the main change in our understanding of the future fate of the universe in the last twenty years. It was thought until about twenty

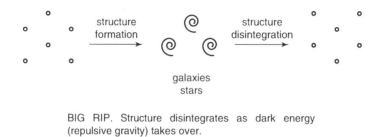

BIG RIP. Structure disintegrates as dark energy (repulsive gravity) takes over.

years ago that the matter content, including just baryons and nonbaryonic dark matter, was the whole story. Dark energy was unknown.

ಹಿ ೀ ೆ ೊ

Let us take a more exotic and theoretically-disfavored case where the dark energy equation of state is more negative than minus one and constant. This is consistent, possibly even favored by the status of the CMB observations, particularly those by the WMAP collaboration.

This means that the dark energy has an energy density which actually grows with time, rather than being a constant as in the previous case. It is not surprising therefore that for this scenario, the dark energy takes over even more rapidly and what happens is that the scale factor gauging the size of the universe becomes infinite after only a finite time.

This blow up in the scale factor needs some elaboration. Firstly, this will not happen any time soon. The earliest could be in about ten billion years. The Earth would have been swallowed as the Sun enters its red giant phase after another five billion years so any life surviving on Earth will have been annihilated long before this so-called Big Rip.

But if there are observational astronomers safely away from the Sun but still in the Milky Way, they will cease to see galaxies outside of our Local Group because they will have expanded away to infinity.

It will be as though the Local Group were the entire universe. Iron-ically, this is similar to the way the universe was perceived in 1917 when the first real theoretical cosmology paper was published.

At that future time, locally-bound systems held together by gravi-tational attraction can continue with their local time but the notion of cosmological time will end. One can justifiably call the phenomenon the end of time. But it is important to distinguish two kinds of time: local and cosmological. Only the cosmological version ends and peo-ple locally can still wear watches as usual, though only temporarily as we shall see. This ending of cosmology is possible only in the case of the equation of state below minus one. Why is this theoretically disfavored?

One assumption typically made in general relativity is that the energy density is not negative in any inertial frame related by special relativity. This is called the weak energy condition and is based on intuition of what is plausible. With such an assumption, an equation of state below minus one is impossible.

Nevertheless, it is a question which must be settled by observation, not by theory. At first some observers of the Type 1A supernovae presented their data assuming the weak energy condition obtained. But they learned that it is better not to listen to theorists and to keep an open mind. Plotting the data with an unconstranied equation of state is much more useful to the theorist. The first data released by the WMAP collaboration studying the CMB made similar assumptions about the weak energy condition but they have now reanalyzed their data without such an assumption.

ॐ ॐ ॐ ॐ

If the dark energy is going to require a revolutionary new theory for its understanding, then the less prejudiced the input the better.

Clearly at risk, especially in this exotic scenario, is classical general relativity itself. What could be more interesting? Of course, such a radical departure is rightly to be resisted and regarded as a last resort by conservative theorists.

But let us pursue this radical possibility even further. The presence of a negative energy density is suggestive of instability and that the dark energy may be able to decay into a lower energy state, for example one where there is no dark energy. If this is what is technically called a first-order phase transition, like water boiling into steam, then the metastable dark energy is analogous to superheated water. Such superheated water is water carefully heated in a very clean vessel to above its boiling point. It may remain in this state for a considerable time but eventually a bubble of steam will exceed a critical size and expand to precipitate boiling of all the water.

The idea of a critical radius also appears for the dark energy but here the critical radius is truly astronomical, of at least galactic size. This means that there is a huge barrier to the decay and in this way one can reconcile the metastability of dark energy with the fact that it has existed already for billions of years.

❧   ❧   ❧   ❧

As has already been mentioned, one further aspect of the super-negative equation of state is that it leads to an unusual future for the universe. The scale factor characterizing the expansion of the universe becomes infinite after a finite future time, maybe just a few hundreds of billions of years.

This may not bother planet Earth too much because after another five billion years or so our Sun will swell up into a red giant as its store of thermonuclear power depletes and engulf the Earth and any life surviving here.

Nevertheless this Big Rip expansion would effect small as well as large objects. We mentioned above that people could wear watches as usual. That was not completely true because, apart from there possibly being no people to wear watches, the dark energy density grows so rapidly towards an infinite value that even stars like the Sun, even if already a red giant, and smaller things like watches and individual atoms will get torn apart by this overwhelmingly repulsive gravitational force generated by dark energy. Whether or not this happens depends on the causal structure of the resultant spacetime which in turn depends on the fluctuation spectrum of the dark energy. Present data are consistent with perfect smoothness of the dark energy but it seems possible that there exists fluctuations for it just as there are for dark matter.

ôð ôð ôð ôð

If the equation of state, on the other hand, is more positive than minus one then the future growth of the dark energy can be less rapid than for a cosmological constant. This is especially so in a situation where the equation of state is time dependent such that it actually becomes positive at some future time. After that time the universe will become dominated by the baryons and nonbaryonic dark matter and will therefore evolve just as in the situation without dark energy.

An equation of state more positive than minus one occurs naturally in a theory with a scalar field which plays a role for dark energy similar to that played for inflation by the scalar inflaton. Such a theory is generically called quintessence. In some versions it may be possible that the same scalar field plays the role of the inflaton and of the quintessence field. This is called quintessential inflation.

There is plenty of freedom in inventing a quintessence model and it is never completely clear whether quintessence is any more than a parametrization of ignorance. Nevertheless, it is a lively topic of research and can give some theoretical framework for comparison with the observational data.

❧ ❧ ❧ ❧

So the fate of the universe is as unknown as the equation of state of the dark energy. The most conservative possibility is that the latter is constant and equals minus one. It is then the cosmological constant introduced by Einstein in 1917 for a completely different reason and with the opposite sign.

❧ ❧ ❧ ❧

This still leaves the severe issue of fine-tuning in that the value of the constant is over one hundred and twenty orders of magnitude smaller than it would naturally appear in a gravitational theory.

More exciting surely, and no less likely, is a radical equation of state more negative than minus one. Such an outcome of observations would lead to a crisis in theoretical physics as severe as the one created by the aether issue over a hundred years ago. It could, by the dictum that necessity is the mother of invention, then lead to dramatic and revolutionary progress in our understanding of the underlying theory.

It used to be thought that the future fate of the universe, before the discovery of the dark energy component, depended simply on whether the matter content Ω_M satisfied $\Omega_M > 1$ which gives a positive curvature closed universe which will stop expansion after a finite time and recontract to a Big Crunch. Or if $\Omega_M < 1$ there is a negative curvature open universe which expands forever. Finally if $\Omega_M = 1$ there is a flat universe which also expands forever, coming to rest at asymptotically infinite positive time.

The present knowledge is that $\Omega_{\text{total}} = 1$ (at least very nearly) and so the universe is flat. This resembles most the $\Omega_M = 1$ flat universe without dark energy. But the dark energy component introduces a major uncertainty, especially with regard to its equation of state. It is possible that this equation of state will never be known sufficiently accurately from observation such that one can extrapolate into the future with absolute confidence. To that extent the future fate of the universe may never be known with certainty.

One other possible future scenario is that the theory itself will become more certain. For example, a successful theory of quantum gravity that explained the past history of the universe could be trusted to predict the future. This could come from string theory which is the most promising candidate for a consistent theory of quantum gravity at this time. On the other hand, there is such a variety of candidate vacua in string theory, possibly 10^{500} or more, that it may require a better understood theory to make a definite prediction.

A successful theory of quantum gravity would be expected to also shed light on the origin of the universe. What happened before inflation? Was inflation inevitable? Was there an eternal repetition of inflationary eras leading to an infinite number of universes, called a multiverse?

Some of these questions edge towards the limits of scientific enquiry and may never (a dangerous word!) be answered with certainty. Some scientists use the principle that our universe must be such as to permit the evolution of intelligent life. This so-called "anthropic principle" can be invoked to account for the values of the dimensionless physical parameters. The majority of people find this device unscientific because it replaces physical explanation by a principle which involves biology.

Still any theory which is to be trusted to predict the future fate of our universe must surely be able to provide information pertaining to its origins. Both questions are valid lines of scientific enquiry. It can and has been argued that the future fate of the universe is not within physics because the prediction cannot be tested. This seems to be only a question of semantics because in principle it can be tested if only one has the patience to wait for a few billion years.

At present we can be sure of our extrapolation back to a temperature of the equivalent of one hundred proton masses which was obtained some ten-billionth of a second after the Big Bang. As high-energy colliders push back the energy frontier this will recede by another order of magnitude or two towards a trillionth of a second after the Big Bang. This is still too late for inflation and evidence for inflation will necessarily be more indirect. Most compelling would be observation of the gravity waves emanating from the inflationary era but for these to be of detectable strength it requires that the characteristic energy scale of inflation be very high, not too far below the Planck energy. It remains to be seen whether Nature chooses such a high inflation energy scale.

かかかか

As we have seen, direct evidence for inflation is particularly elusive simply because, if it occurred, it did so only in the first billionth of a second, possibly even in the first trillion-trillion-trillionth of a second after the Big Bang. Electromagnetic radiation measurements such as optical or radio telescopes are sensitive only back to the surface of last scatter some hundred thousand years later. Abundances of helium and hydrogen and other light isotopes can tell us indirectly about one minute after inflation, which is still a very long time by early universe standards. Neutrinos may tell us directly about the same period.

The only possible direct evidence for inflation would seem to be made possible by detecting weak gravity waves emanating from that era, especially if the inflation takes place extremely early on. This is an exciting possibility for which we will have to wait for at least a few more years.

ॐ ॐ ॐ ॐ

The discussions of the fate of the universe can be criticized as untestable because it will be many billions of years before the actual event. It has been argued that the discussion of such matters is not even physics, or science.

This seems unnecessarily semantic since if a theory is successful in accommodating what is known about the past it is very natural to ask what it predicts for the very distant future. The time scales involved in cosmology are typically gigantic compared to the human lifetime. But in physics, especially cosmology, there is nothing significant about the biological scale of a human lifetime. It is equally valid to ask what happens to the universe in the next hundred billion years as to enquire about the decay of an unstable particle in a tiny fraction of a second. There is no difference in principle.

The separation of physics from biology is called into question also in the use of the anthropic principle: that the universe must be such as to allow intelligent life which can observe it. The use of this principle is quite controversial since it is a way of avoiding a physical explanation for certain aspects and parameters describing our universe.

While it is true that without intelligent observers the science of cosmology would be impossible, it seems to be a "cop out" on scientific explanation to appeal to such an idea to, for example, say that the lifetime of the universe must be above some minimal value or that the couplings and parameters in the underlying fundamental theory must be within small ranges around their observed values otherwise the cosmic evolution would be so different as to disallow the creation of life.

So the preferable path is to study physics without the need to input facts from biology with the idea that the occurrence of intelligent life is not of central importance to the physics rather than constraining the physics such that life is possible. This attitude is more likely to lead to explanations which are satisfying.

As we have seen, in the minimal Big Bang scenario, the previous history of the universe is finite with an age of 13.7 billion years. In some future scenarios, particularly where the dark energy has an equation of state less than minus one, the cosmos ends at a finite time in the future when the scale factor diverges to infinity. This would provide a philosophically pleasing symmetry between the future and the past.

One possibility, for which there exists no substantial evidence but which has some theoretical appeal, is that the universe is finite also spatially. This can occur if the universe has non-trivial topology in space. If this were the case, it would provide an even more symmetric view between space and time.

ॐ ॐ ॐ ॐ

The recent growth in knowledge about the universe has been astonishing, as it usually happens, but answering several questions gives rise to others. At present there are great enigmas about the universe. Perhaps the most striking mysteries are: what is the dark matter? what is the dark energy?

The existence of such enigmas is very healthy for research in the field because they signal the distinct possibility that revolutionary advances in the theory, as equally dramatic as relativity and quantum mechanics were in the previous century, will be necessary. When old ideas are tried and fail it is all the more likely that the intelligence and creativity of theorists are the only avenue to advance knowledge and understanding.

❧ CHAPTER 7 ❧

Advantages of Cyclic Cosmology

We have, by now, discussed how the past and the future lifetimes of the universe may be finite by virtue of the Big Bang and the Big Rip respectively. The Big Bang was first discussed in the 1920s and some physicists found it disturbing that the density and temperature of the universe become infinite at a finite time (now known as 13.7 billion years) in the past.

This concern led several of the leading scientists to explore an alternative where the universe cycles between expansion to a turnaround before contracting to a bounce, such that the temperature and density always remain finite. This happens an infinite number of times, thus eliminating any start or end of time.

However, the theorists in the 1920s and 1930s could not construct a model with this cyclic property. The principal obstruction was entropy and the second law of thermodynamics. During each cycle the entropy necessarily increases so that subsequent cycles become longer and expand to a larger size. Extrapolating back in time the cycles become shorter and smaller until eventually one arrives at a Big Bang again with the same issues of initial conditions as in the non-cyclic universe. This property removes the motivation to study the cyclic scenario.

As we have discussed at length in the two previous chapters, the accelerated expansion of the universe discovered only in 1998 requires a large fraction, or over seventy percent, of the Universe to be in the form of dark energy. This is something we now know which the theorists of old (Friedmann, LeMaitre, De Sitter, Einstein, Tolman) did not know. Therefore, we may consider whether dark energy can help overcome the obstruction which entropy provided to oppose construction of a consistent cyclic cosmology which could preclude a start and an end.

To understand how this problem is addressed, we need to explain the concept of entropy. Consider a room full of air molecules. The number of molecules is extermely large, typically of order 10^{27} or one thousand trillion trillion. Written out this is 1,000,000,000,000,000,000,000,000,000 molecules! These molecules are moving with high velocity, typically about 300 meters per second at room temperature and constantly colliding with one another.

One might think that describing such a system is hopeless, which it is in detail such as for the motion of any individual molecule. But the very large numbers lead to simple laws for the entire system. This is the physics of thermodynamics and statistical mechanics of which a principal architect was Boltzmann.

෨ ෯ ෨ ෯

Let us think about the following question. Air molecules are about twenty percent oxygen and eighty percent nitrogen. Oxygen is crucial to us to breathe and survive: without any we will pass out within a few minutes, and then die. How do we know that the volume surrounding our nose and mouth will not be depleted of all oxygen molecules for a few minutes? Why are there not ambulances waiting for just such an emergency? Answering this question will introduce the concepts of entropy and the second law of thermodynamcs which play a central role in cyclic cosmology.

The molecules is a room can be in an astronomical number of configurations with regard to the positions and velocities of all the individual molecules. Statistical mechanics is the study of such configurations and its basic assumption is that every configuration is equally likely. The system will quickly evolve into a configuration which corresponds to the maximum probability. Such configurations are those of highest entropy which is a measure of the disorder of the system. The highest entropy states are favored by an enormous factor.

For such an equilibrium state of highest entropy one can calculate the velocity distribution of the molecules and the temperature to extreme precision because of the very high statistics involved. Coming back to our question, the configuration with no oxygen molecules in the vicinity of a person's nose and mouth for a few minutes is a highly ordered state of lower entropy and of extremely low probability. Boltzmann's second law of thermodynamics (the first and the third are not relevant here) states that the entropy of an isolated system always remains constant or increases; entropy never decreases. Thus a room full of air molecules will never lower the entropy to this unlikely configuration.

We can calculate the probability of this happening using statistical mechanics. The result is the following, as I have sometimes asked to a class: if I write the probability as 0.0000000… on the blackboard with each zero occupying one inch where will be the first non-zero digit? Will it be at the end of the blackboard? At the other side of campus? A hundred miles away? The amazing answer is that it will not be in the visible universe so no ambulances are necessary.

Transition to such a configuration would violate the second law of thermodynamics because the entropy would decrease. It is important, however, to emphasize that such an unlikely configuration would not violate any fundamental law of physics in the microscopic regime. The second law is only a statistical law, yet because of the huge number of molecules the statistical law might as well be an exact one for practical purposes. Historically it was this subtle distinction bewtween an exact law which is never violated and a statistical law which is *practically never* violated for a system of a million trillion trillion molecules that led to only a slow acceptance of Boltzmann's profound result, especially by mathematicians, and possibly contributing to his eventual suicide.

Another central aspect of statistical mechanics is provided by phase transitions such as from ice to water or water to steam. The thermodynamics of phase transitions was first systematized in a general treatment of statistical mechanics by one of the earliest American professors of theoretical physics, Gibbs.

❧ ❧ ❧ ❧

The ideas of entropy and the second law can be applied to the CMB as a whole. The entropy of the present universe can be estimated as $\sim 10^{88}$ which is a one followed by eighty-eight zeros. This entropy lies not only in the radiation, also in the matter both dark and luminous. As the universe expands the entropy necessarily increases according to the second law. The entropy of the radiation component remains constant and we say that it expands adiabatically; the entropy associated with matter gradually increases as a result of irreversible processes.

In the very early universe, entropy increased during inflation by a factor of $\sim 10^{84}$, so comparison with the present entropy reveals that the radiation entropy at the beginning of inflation was extremely low, $\sim 10^4$. This last number is essentially zero on the scale of the later entropies. In our discussion, we will come across two magnitudes of the entropy: $\geq 10^{88}$ will be called *large* entropy, and $\leq 10^4$ will be called *small* or *essentially zero* entropy. As can be seen, a cyclic universe with periodic entropy must involve a dramatic *decrease* in entropy to compensate for the huge increase in entropy at inflation, but how can any decrease be consistent with the second law of thermodynamics which demands increasing entropy?

This is the question that stymied progress toward an oscillating universe in the 1920s and 1930s. But why did the cyclic universe seem an attractive alternative to the non-cyclic scenario?

Once it was realized by Friedman in 1922 and independently by Lemaître in 1927 that the equations of general relativity led naturally to an expanding universe, not a quasi-static one as originally proposed by Einstein in 1917, it was also seen that there was one undesirable feature. Namely, as one extrapolates into the past at a finite time (now known as 13.7 billion years) the temperature and density become infinite, and the scale factor reduces to zero. This leads to the idea of the explosion of a primordial "atom" or a Big Bang at the initial time $t = 0$. But the problem is that the classical equations cease to be applicable at this singularity, and general relativity cannot hold.

One reponse is that at a sufficiently early time, the Planck time, $\sim 10^{-44}$ s after the Big Bang, the effects of quantum mechanics must enter so that, in any case, the classical Friedman equations fail and singular behavior at $t = 0$ with infinite density and temperature and vanishing scale factor could be avoided if we knew a satisfactory theory of quantum gravity. Such a complete theory is unknown but in attempts at such quantum cosmology various attempts have been to formulate satisfactory initial conditions for the quantum version of the Big Bang. Without entering into technicalities, such ideas that are the most studied are due to Vilenkin, and to Hartle and Hawking. It is probably fair to say that the jury is still out on these possible alternative attempts to circumvent the Big Bang singularity.

At the classical level the Friedman equations make a hypothesis known as the Cosmological Principle. This has two components: (i) the universe is assumed to be perfectly homogeneous, and (ii) it is assumed to be perfectly isotropic or rotationally symmetric. Now it is clear that neither component of the Cosmological Principle is exactly valid at smaller scales such as the size of galaxies and clusters of galaxies but at extremely large scales, an order of magnitude above cluster sizes, it does seem to be a good approximation.

Nevertheless, a legitimate question for classical general relativity is whether relaxing either homogeneity or isotropy or both could avoid the Big Bang singularity? An answer to this question was offered in the 1960s by Hawking and Penrose who showed under general conditions that a past singularity was inevitable and did not depend on the Cosmological Principle.

However, the Hawking and Penrose no-go theorem necessarily made assumptions, one of which will be seen to be important. They assumed the physically plausible requirement that the energy density must never be negative, since such an energy density seemed to have no physical interpretation. As we shall see, it is this assumption that must be avoided in making a workable cyclic cosmology.

∽ ∽ ∽ ∽

Going back to the 1920s, it seemed desirable to avoid the Big Bang singularity even classically. One attractive possibility was a cyclic theory in which the density and temperature always remain finite and the scale factor is never zero. We have already discussed

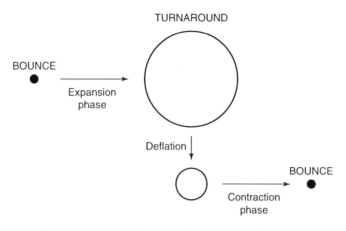

CYCLIC UNIVERSE: Stages of bounce, expansion phase, turnaround (deflation), contraction phase back to bounce.

the independent attempts by Friedman and Lemaître respectively in 1922 and 1927.

It is significant that when he heard a seminar at California Institute of Technology in 1931 by Lemaître on his research, Einstein was very enthusiastic in his reception of the cyclic ideas. De Sitter and Einstein published jointly on the topic in 1932.

A particularly clear and endearingly modest discussion of the role of entropy in cyclic cosmology is found in the book *Relativity, Thermodynamics and Cosmology* by Tolman.

To add a personal anecdote, when I was studying for the Final Honor School in Oxford in 1965 and had immediate access to several hundreds of physics books, my personal favorite book was always Tolman's. I do recall spending hours then intrigued by the apparent contradiction between the attractive idea of cyclic cosmology and the second law of thermodynamics; the contents of Tolman's book, however, did not appear in my examinations.

ॐ ॐ ॐ ॐ

One important fact about the universe, discovered only in 1998, and obviously unknown in the 1920s and 1930s, is that the expansion is actually accelerating. As we have previously discussed, this led to the identification of over seventy percent of the energy as dark energy. This then led to our question of whether dark energy can intercede in the apparent contradiction?

Before explaining a positive response to this question, let us make some general considerations of cyclicity and entropy in an attempt to make the solution seem inevitable.

A cyclic universe goes through four stages: Bounce $\rightarrow$ Expansion $\rightarrow$ Turnaround $\rightarrow$ Contraction $\rightarrow$ Bounce, repeated an infinite number of times.

During expansion, entropy initially increases by an enormous factor $\sim 10^{84}$ during inflation and then more gradually increases due to irreversible processes in the matter component. We recall that the radiation expands with constant entropy and the dark energy has zero entropy.

Before discussing the turnaround, consider the contraction phase which presents its own peculiar issues. One is that matter and dust will form structure more readily than during expansion. Black holes, if present, will expand and more will form, further impeding any smooth contraction.

$\approx$ $\approx$ $\approx$ $\approx$

There is an even more serious problem with matter content during contraction, namely that several phase transitions must take place *in reverse* to proceed successfully back to a bounce in the early universe. For example the phenomenon of recombination would require, in reverse, that neutral atoms dissociate into protons and electrons as the temperature increase. This would decrease entropy and be statistically impossible as a violation of the second law of thermodynamics.

అం అ అ అ

There are other phase transitions such as the so-called quantum chromodynamics transition at a few hundred MeV where quarks and gluons become confined into hadrons. Also, there is the weak transition at about one hundred thousand MeV at which electroweak forces separate into weak and electromagnetic forces.

How can a contracting universe pass in reverse through such phase transitions without violating the second law of thermodynamics?

One possibility which is seductive but which we shall quickly discard is that during contraction the "arrow of time" reverses. This is an assumption made to argue that entropy may decrease during the contracting phase.

What does it mean? The arrow of time refers to the second law of thermodynamics of a statistical mechanical system for which increase of entropy defines a "forward" direction of time. At a biological level, we remember the past but not the future, we also grow older so for us there is surely a well-defined arrow of time. For statistical systems in physics the arrow of time is equally well defined by the behavior of the entropy.

The entropy problem could be solved very simply if we were allowed to adopt a reversal of the arrow of time during contraction and entropy could then decrease. If we consider this further, however, it is only a semantic device corresponding to unacceptable physics although it has been used by some authors (including, for a while, by this one!) as a last resort to address the thorny cyclicity problem. In a universe with a reversed arrow of time, statistical mechanics would become nonsensical because equilibrium states of, say, an ideal gas would proceed to disequilibriate into very unlikely configurations, which is evidently nonsensical.

Reversal of the arrow of time is therefore an absurdity better to avoid if at all possible. Fortunately we shall find a solution to cyclicity in which the arrow of time keeps moving forward in the conventional sense.

ॐ ॐ ॐ ॐ

To proceed towards an acceptable solution, we need to introduce a couple of new concepts: branes and causal patches.

ॐ ॐ ॐ ॐ

Branes were first emphasized in the 1990s on the basis of string theory, particularly by Polchinski. String theory was believed in the 1980s to be a theory of one-dimensional extended objects, namely strings, interacting with one another. As such, it was already a remarkably consistent generalization of quantum field theory which included within it both gauge field theory and general relativity. Moreover, it provided a finite theory of quantum gravity for the first time.

Around 1990, however, it was shown that a theory of only one-dimensional strings is not internally consistent and must include higher dimensional "branes", a shortened version of the word "membranes". In this language, strings are 1-branes. In a ten dimensional superstring theory branes appear in addition as p-branes for all p between 2 and 9. Which values of p appear depends on the particular superstring considered; for example, a 9-brane in 10 spacetimes dimensions is a space-filling brane.

Of special interest to cosmology is the 3-brane. This notion has by now been abstracted from string theory leading to the idea of a brane world in which non-gravitational physics is restricted to the 3-brane on which we are assumed to live. Strong and electroweak interactions are restricted to this "TeV" brane, while gravity alone spreads also into the extra spatial dimension or dimensions in which the TeV 3-brane happens to be embedded. In the theory of brane worlds one remains agnostic about the number of extra space dimensions which can vary from one up to the six as predicted by superstrings or seven as allowed in M theory. In our discussion we shall, for simplicity, assume only one extra space dimension, or a total of five spacetime dimensions.

One popular version of the brane world involves two parallel 3-branes where one is the TeV 3-brane and the second is a "Planck" 3-brane from which gravity originates. This theory offers an attractive explanation of why gravitation is so weak in our observed world. It is remarkable, for example, that the electric force between two protons separated by, say, an atomic size (or any other separation) is some forty orders of magnitude larger than the gravitational attraction; that is a one followed by forty zeros. Another reflection of the weakness of gravity is that a small magnet will hold an object to a refrigerator door against the gravitational pull of the entire Earth.

For cosmology, our universe will be the TeV brane so in such a brane world one can derive a Friedman equation for this TeV brane which has a crucial modification. Without going into technical details, a new term appears which involves a parameter with the dimension of a density and this critical density signals when the expansion of the universe will stop and turnaround into a contraction mode. This same critical density determines what will be the bounce temperature at which the contracting universe will stop and bounce into a new expanding mode.

The critical density is related to the brane tension and the new brane term in the Friedman equation plays a central role in our solution for the cyclic universe.

ॐ ॐ ॐ ॐ

A central role is also played equally by the concept of causal patch.

ॐ ॐ ॐ ॐ

To introduce the causal patch we must revisit the Big Rip where the present expansion of the universe ends at a finite time in the future when the universe rips itself apart and the dark energy density as well as the scale factor diverges to infinity.

As one approaches the Big Rip, the structures in the universe successively disintegrate beginning with the largest scales. An approximate rule is that once the dark energy density reaches the average density of a gravitationally bound system, that system will become unbound. This implies that first the clusters of galaxies disintegrate followed by the galaxies and then the solar system. Nothing is immune to this disintegration process and eventually even atoms, atomic nuclei and nucleons — protons and neutrons — will become unbound. These are just the smallest systems we currently know and if there exist yet smaller higher density bound sytems, they too would be subject to the unbinding process.

As another example which is important for the sequel, black holes are themselves torn apart on the approach to the Big Rip.

ॐ ॐ ॐ ॐ

At a time somewhat later than the unbinding of a system, the bound components become causally disconnected, meaning that they cannot communicate even at the speed of light before the universe ends. Eventually we may regard the universe itself as disintegrating into a huge number $> 10^{103}$ of causal patches which are disjoint and separate. The idea now is to delay the brane induced turnaround until a trillion trillionth of a second or less before the would be Rip. At this time, one causal patch contains no quarks or leptons and certainly no black holes.

The causal patch at the turnaround contains only dark energy and a vanishingly small number of photons. Its entropy is *small* in the sense discussed earlier, meaning it may have entropy equal to zero. Before the unbinding process, on the other hand, the entropy was *large* by the same discussion, meaning it was at least a one followed by over one hundred zeros.

This is essentially the reverse of the vast increase of entropy experienced during inflation and so may naturally be called deflation.

怘 怙 怠 怡

We have now assembled all the pieces of our proposed cyclic universe. It will involve ending the present expansion at a finite time in the future, typically of the order of a trillion years if the equation of state of the dark energy is just below minus one say, -1.01.

At the turnaround, only one causal patch will be retained as our universe and it contracts with constant *small* entropy: the radiation of any photons contracts with no increase of entropy and the dark energy has zero entropy.

What is equally important is that the contracting universe is empty of all matter, whether dust or black holes, and thus can contract without confronting all the difficulties discussed earlier such as enhanced structure formation and growth of black holes as well as formation of new black holes.

The contracting universe, much smaller than its expanding pregenitor, continues until the dark energy again reaches the critical value. Then the contraction ceases at a certain bounce temperature and the universe cycles into inflationary expansion again with the entropy increasing from a *small* to a *large* value. The temperature at which this happens is related to the critical density in the model.

ॐ ॐ ॐ ॐ

This provides a possible model for a cyclic universe which respects the second law of thermodynamics.

It has two ingredients which could not have been foreseen in the 1920s and 1930s by the theorists who attempted such a model. The first is dark energy which was deduced from the observed accelerated expansion rate first time in 1998. The second is the idea of a higher dimensional brane world which has emerged in theoretical physics just as recently.

ॐ ॐ ॐ ॐ

Another hurdle of more recent vintage was the singularity theorems of Hawking and Penrose from the 1960s. These assumed an everywhere positive energy density while a dark energy with equation of state less than minus one violates such an assumption.

ॐ ॐ ॐ ॐ

Only time will tell whether the cyclic model we have discussed will survive closer scrutiny. A general observation, however, is that a cyclic universe which avoids classically any singular or infinite behavior in density or temperature, and which avoids any beginning or end of time, seems more aesthetically acceptable than a cosmology which originates from a singularity.

ॐ ॐ ॐ ॐ

This new viewpoint on the future of the universe is possible only given the amazing developments in both theory and observations since the very end of the twentieth century. The conclusion is that conventional cosmology with time starting at the Big Bang and continuing for ever in an infinite expansion now has plausible alternatives, especially in this cyclic model where time never begins or ends.

Summary of Answers to the Questions: Did Time Begin? Will Time End?

There are three different futures for the universe which we have discussed. One key question is whether the present expansion phase will continue for an infinite time which is conventional wisdom, or will it stop after a finite time of order a trillion years? If it stops, a second question is whether our universe will then contract and bounce cyclically?

٫ ٤ ٫ ٤

Although we cannot say with certainty which future is correct, progress in addressing this question has been so rapid that it is possible to order these three futures, according only to their aesthetics, in *decreasing* probability with conventional wisdom third.

$$\text{۞ ۞ ۞ ۞}$$

The scenarios in (1) and (2) assume the equation of state of dark energy ω is slightly below -1, say $\omega = -1.01$, while scenario (3) adopts $\omega = -1$ exactly.

(1) **Most likely.** The present expansion will end after a finite time, the universe will contract, bounce and repeat the cycle. A cyclic universe. *Time had no beginning and will have no end.* This presumes that the entropy problem has been resolved as discussed in this book.

(2) **Next most likely.** The present expansion will end after a finite time in a Big Rip. *Time began* in the Big Bang some 13.7 billion years ago *and will end* some trillion years in the future.

(3) **Least likely.** The present expansion will continue for an infinite time as for a cosmological constant. *Time began* 13.7 billion years ago *and will never end.* This is the prevailing conventional wisdom which is here being challenged.

Glossary

Aether

A hypothesized, but later discarded, medium through which **electromagnetic radiation** propagates.

Baryonic matter

Matter, including the stuff we and all everyday things are made of, is comprised by mass almost entirely of baryons. Baryonic matter forms about 4% of the energy of the universe.

Big Bang

By using the **Friedmann equation** and the known contributions of matter and radiation it is possible to conclude that temperature and density became infinite some 13.7 billion years ago at a Big Bang when time possibly began.

Big Rip

If the **dark energy** has sufficiently negative pressure it leads to an end of time before which the structure of the universe is ripped apart.

Bounce
The transition from contraction to expansion in cyclic cosmology.

Branes
Multidimensional objects which occur in **string theory**.

CERN
European laboratory for high energy physics near Geneva, Switzerland and home of the LHC.

COBE
COsmic Background Explorer. This satellite was, in 1992, first to detect the anisotropy of the CMB.

CMB
Cosmic Microwave Background of photons propagating from the **surface of last scatter** at **recombination**.

Cosmological principle
The assumption that the universe is **homogeneous** and **isotropic**.

Cosmology
The scientific study of the universe.

Critical density
A special value of mean energy density of the universe which implies that the geometry of the universe is Euclidean.

Cyclic cosmology
A theory which avoids infinite temperature and density by infinite repetition of the sequence **expansion-turnaround-contraction-bounce**.

Dark energy
The accelerating cosmological expansion discovered in 1998 necessitates this component of energy with negative pressure and comprising 72% of total energy.

Dark matter
About 24% of the total energy of the universe comprising primarily **nonbaryonic dark matter**.

Doppler effect
The frequency of a wave which may be sound or light is effected by the relative motion of the source and observer. This effect was first discovered by Doppler in 1845. In a dramatic demonstration of his discovery, he arranged for trumpeters to play on an open wagon of a moving train.

Earth
Third nearest planet to the **Sun**.

Electromagnetic radiation
Oscillating and propagating electric and magnetic fields exemplified by the **CMB**.

Friedmann equation
Governs the time dependence of the **scale factor** of the universe and is derived from **general relativity** and the **cosmological principle**.

General Relativity
Theory which relates geometry of spacetime to mass distribution.

Helium
Second lightest element, about one quarter by mass of the universe.

Homogeneous
Uniformly distributed, the same at every position.

Hubble Law
Recession velocity is proportional to distance.

Hubble parameter
Ratio of recession speed to distance. Its units are kilometers per second per megaparsec.

Hubble Space Telescope
An orbiting instrument with a 2.4 meter primary mirror deployed by NASA astronauts in 1990.

Hydrogen
Lightest element and three quarters by mass of the universe.

Inflation
Period of very rapid expansion hypothesized in the first trillionth of a second, or less, of the expansion era.

Inflaton
Hypothesized particle underlying **inflation**.

Isotropic
The same in all directions, also known as rotational invariance.

Jupiter
The fifth planet from the **Sun** and the largest planet in the **Solar System**. The largest four of its moons were first observed in 1610 by Galileo using an early telescope.

LHC
Large Hadron Collider. It is expected to start producing data from proton–proton collisions at unprecedented high energy in 2010.

MACHO
MAssive Compact Halo Object, a possible constituent for **dark matter**.

Mars
The fourth planet from the **Sun**.

Mercury
Nearest planet to the **Sun**.

Milky Way
Galaxy containing the **Solar System**.

Moon
Satellite of the **Earth**.

Naturalness
Criterion that a theory should not contain unexplained large dimensionless numbers.

Neutrino
Elementary particle which experiences only the weak interaction.

Nonbaryonic dark matter
Contributes 24% of the energy of the universe. Not baryons and not luminous.

Nontrivial topology
Idea that space has special properties at the largest distances. There is no evidence for this.

Photon
Massless elementary particle which is the smallest unit of **electromagnetic radiation**.

Principia
Book by Newton published in 1687. For the next 200 years, this book, now of only historical interest, had significant influence. It contained his three laws of mechanics and his law of universal gravity. Although the author invented calculus, he remarkably avoided using it in the book even though this made some of his proofs more challenging.

Quantum mechanics
The theory invented in 1925 which explains spectra and the stability of atoms, as well as many other hitherto profound mysteries such as the distinction between conductors and insulators.

Recombination
Cosmological era when charged particles bind to form neutral atoms and the universe becomes transparent to **electromagnetic radiation**. It is a misnomer because the particles were not previously combined.

Saturn
The sixth planet from the **Sun** and the second largest planet in the **Solar System**.

Scale factor
Measures distance between galaxies and obeys the **Friedmann equation.**

Scientific method
Systematic making of experimental observations, framing hypotheses and theories which make predictions followed by further experimental tests.

Solar System
The **Sun** and its eight planets.

Special Relativity
Theory which accommodates motion at speeds comparable to the speed of light, proposed by Einstein in 1905 based on a symmetry of Maxwell's equations which he generalized.

Steady-state theory
A once popular, now discredited, alternative to the **Big Bang**.

String theory
A theory based on an assumption that the fundamental entities are not point particles but extended objects.

Sun
The nearest star to the **Earth**. Its present age is about five billion years and it is expected to run out of hydrogen fuel in another similar time. Our understanding of the sun and its evolution is remarkably complete. The second nearest star is a million times further away and harder to study.

Surface of last scatter
An opaque wall corresponding to **recombination** where the **CMB** photons originate.

T-duality
Symmetry of **string theory** which relates small and large distances and can play a role in applications to **cosmology**.

Turnaround
Transition from expansion to contraction eras in **cyclic cosmology**.

Uranus
The seventh planet from the **Sun**.

Venus
The second planet from the **Sun**.

WIMP
Weakly Interacting Massive Particle, a candidate for a dark matter particle.

WMAP
Wilkinson Microwave Anisotropy Probe. Its first data released in 2003 gave values of cosmological parameters with unprecedented high accuracy.